GENETICS - RESEARCH AND ISSUES

CORPUSCULAR-WAVE NATURE AND WAVE PROPERTIES OF PLANT CELLS

GENETICS - RESEARCH AND ISSUES

Additional books in this series can be found on Nova's website
under the Series tab.

Additional e-books in this series can be found on Nova's website
under the e-book tab.

GENETICS - RESEARCH AND ISSUES

CORPUSCULAR-WAVE NATURE AND WAVE PROPERTIES OF PLANT CELLS

NINA F. KUZNETSOVA

NOVA BIOMEDICAL

New York

For permission to use material from this book please contact us:
Telephone 631-231-7269; Fax 631-231-8175
Web Site: http://www.novapublishers.com

NOTICE TO THE READER

Additional color graphics may be available in the e-book version of this book.

Library of Congress Cataloging-in-Publication Data

ISBN: 978-1-62618-785-6

Library of Congress Control Number: 2013939132

Published by Nova Science Publishers, Inc. † New York

Contents

Preface

Quantum biology is a wide area of research closely connected with almost all parts of biology. It is based on experimental data of biological sciences and the fundamental laws of physics (de Broglie law of corpuscular-wave dualism of the matter, the conservation laws, including the laws of thermodynamics). At this time, our knowledge in this area is fragmentary. The usual corpuscular biology studies only one plane of living matter organization, the structure and function of which is determined by the DNA-particle. That is why the theory often does not agree with experience, the physics laws don't work. It leads to frequent changes of concepts. Many phenomena (division of living matter into cells, restoration and loss of totipotency of cell systems, etc.) do not find an explanation within the corpuscular theory framework.

This book includes 7 chapters. In Chapter I the insight of a cell as a quantum-mechanical system, an equilibrium system, an open and closed system; the notion of biological harmonic oscillator, as an elementary and indivisible unity of the wave properties of a living matter; the principle and regimes of oscillator work in plants; two internal energy sources and their physical nature; the role of DNA-particles and DNA-wave at different hierarchical levels of living matter organization are discussed. In Chapter II the changes of DNA particles, DNA-waves, the cell physical state, its basic components and physiological functions are analyzed during cell cycle of proliferating plant cell. In Chapter III the corpuscular-wave models of nine types of cell division (mitosis, asymmetric mitosis, activating mitosis, meiosis, activating meiosis, free-nucleus mitosis, endomitosis, cleavage and promitosis) are described. The dependence of the principle of prokaryotic and eukaryotic cell development from its condition is shown in Chapter IV. In Chapter V the hypotheses of physical nature of some fundamental processes of biology (separation of organic peace on a plants and animals; differentiation and evolution of sexual gametes by a sex; etc.) are considered. In Chapter VI the mechanisms and ways of energy protection and the reasons for their change during evolution of plants are discussed. How and why a DNA-particle and a DNA-wave change during reproductive development of generative cells is described on *Pinus sylvestris* L. example in Chapter VII.

Chapter I

Introduction to the Problem

Abstract

In this chapter the wave properties and corpuscular-wave model of a plant cell is analyzed. The model is obtained by the synthesis of the quantum and cell theory and the data on the cell biology, molecular biology and developmental biology of plants. Two theoretical positions are prioritized on this basis: 1) the principle of the corpuscular-wave organization of living matter (DNA, chromosomes, cells, etc.); 2) the notion that a cell is a quantum-mechanical system. An elementary unit of the wave properties of cell is determined by the biological harmonious oscillator which functions in a cell as a physical and biological body simultaneously. As the oscillator in eukaryotes is a nucleus together with an oscillating DNA, in prokaryotes it is a cell together with an oscillating DNA. These two sources of inner energy are described, which together with the outer sources (photosynthesis and respiration) provide the cell energetics during the cell cycle. The first source is a biological harmonious oscillator which is present in all types of prokaryotic and eukaryotic cells. The second source is a metaphase chromosome, which meets only in proliferous eukaryotic cells. We describe how a wave function on a DNA molecule in a telophase is formed. The distinctions between a work of biological and physical harmonious oscillators are considered. It is shown, why a cell life as a quantum-mechanical system depends on such physical factors as light, temperature and pressure and the role of DNA-particles and DNA-waves at different hierarchical levels of living matter organization are discussed.

The creation of the cell theory (Huger, 1959) was a result of a knowledge generalization in biology at the end of the 19th century. Its further development proceeded on the basis of genetics, cytology, biochemistry, molecular biology, etc. The achievements in these sciences have considerably expanded the notions of a cell and have established that living matter at the cellular, subcellular and molecular levels is uniform in structure, composition and function. Nevertheless until now a number of the phenomena have found no explanation. This is a list of some of them. 1) Why is the living matter divided into cells? 2) What force initiates cell divisions? 3) Where is the energy for cell doubling concentrated? 4) What is the reason for the aging of cells and their death? And above all, the existing model is not consistent with the fundamental laws of physics: de Broglie law of corpuscular-wave dualism on matter, the laws of thermodynamics and, as a consequence, the law of full energy conservation (Bauer, 1935; Lehninger, 1965, 1985; Swanson and Webster, 1977; Rubin, 1999; Allen, 2008; etc.).

In this chapter the following questions are suggested for discussion: 1) in which moment of a cell's cycle and how the wave function on the DNA molecule can be formed; 2) the principle and operating modes of the harmonic oscillator in a cell; 3) the sources of the cell's inner energy and its physical nature, and 4) the role of the DNA-particle and DNA-wave at different hierarchic levels of the living matter organization.

Formation of a Wave Function on the DNA

It is known that DNA has a dipolar structure (Haggis et al., 1965; Volkenstein, 1965, 1983; Cantor and Scimmel, 1980; Allen, 2008; etc.). Inside each of its nucleotides, the centers of positive and negative charges do not coincide. At the beginning of the telophase the close-packed metaphase chromosomes begin to uncoil. While uncoiling the macromolecule rapidly elongates, and its numerous positive and negative centers come to motion. Their motion under the laws of physics generates the electromagnetic field decreasing as $1/r^2$. The field causes the centripetal motion of charges (ions, molecular dipoles) within the radius of its action. Therefore in the telophase the molecules of proteins, RNA, phospholipids, etc. are drawn to the cell center (Brachet, 1957; Prescott, 1970; Strochkova and Kazan'ev, 1975).

The molecules of phospholipids are asymmetrical by nature: their polar heads are hydrophilic and possess a constant dipole moment, whereas non-polar tails are hydrophobic and neutral (Haggis et al., 1965; Swanson and Webster, 1977; Alberts et al., 2002; Allen, 2008; etc.). When mixed in water the phospholipids are linked in a definite manner. The polar ends of the molecules in a membrane are turned outside, whereas non-polar ones are turned inside. The field concentrates and orients the molecules of phospholipids and their associations along power lines. As they are united into a lipid bilayer, the nuclear envelope is reconstructed. At the moment of its formation the chromosomes, continuing to uncoil and extend, hit the walls of the nucleusand each chromosome during the interphase period occupies its niche in the cavity of the nucleus (Cremer et al., 2000; Chentsov, 2004). Since at decondensation the chromosomes do not change anaphasic orientation, they hit the centromere and two telomere regions. From the impulses created by that contact with the chromosomes, two sinusoidal waves arise, which after meeting each other, give rise to a set of standing waves (chromosomes-waves). Thus, on the assumption of the electrochemical structure of DNA and the mitosis mechanism, in the interphase period the chromosomes should exist in the form of both particles and waves simultaneously.

At the formation of a standing wave on a macromolecule, the cell's environment changes completely. Now we will deal with the oscillating dipoles of chromosomes. During their oscillations the moving charges create the oscillating electromagnetic field, having two zones of action: the close-range (changing as $1/r^2$, $1/r^3$) and far-range (changing as $1/r$) ones. At first this field was described by Gurvich (1944), who called it a biological field. By his data, the biofield acts both inside and outside the cell. As a vector quantity, it defines the order of cytoplasm structure and participates in the uniting of the cells with each other. The physical nature, extent and time of existence of the biofield predetermine the participation of its forces in many processes, such as an active transport of substances through a membrane, the osmotic phenomena, etc.

Cell Is an Open and Closed System

The wave function on the DNA and the biofield exists only in the interphase period, while the nuclear envelope is intact. In this period the cell is an open system. Between the cell and the environment there exists an exchange of energy and matter by means of photosynthesis, respiration, active water transfer, ions and molecules. The nuclear envelope is destroyed during mitosis and with its destruction the wave on the DNA disappears, and chromosomes cease to be a source of the biofield.

It is known that the cell, which started the mitosis, always finishes it, and the order of mitosis phases never varies (Mazia, 1961), i.e., the division mechanism is irreversible and rather independent. Therefore the dividing cell can be viewed as a closed system. By closed system we mean a physical body, which is not affected by outer forces, but within which there are no restrictions regarding inner forces. As such, cell molecules can interact in any way, but the full energy level, according to the first law of thermodynamics, stays constant. At the end of division, when the DNA-wave and biofield will be restored, the cell becomes an open system again. Since the interphase takes about 90 % of a cell's cycle (Mazia, 1961; Swanson and Webster, 1977; Chentsov, 2004), for most of its life the cell exists as an open system.

Biological Harmonic Oscillator Is an Elementary Unit of Wave Properties of the Cell

In the interphase the macromolecule oscillations occur inside the nucleus or the closed cavity with oscillating waves. Hence, in the nucleus, we deal with the waves enclosed in the limited volume, if this is so, then the biological system – the nucleus together with the oscillating DNA – is a physical system – the classical harmonic oscillator, whose work is described by the laws of quantum physics. This harmonic oscillator, which we call a biological system, is an elementary and indivisible unit for defining the wave properties of a cell, and no matter how high the level of development of the living matter is, its biological harmonic oscillator is both a physical and biological body simultaneously. Its work obeys the fundamental laws of physics, and the wave specificity affects the biology of the cell and plant as a whole.

According to quantum physics the biological harmonic oscillator, as well as any oscillating system, should possess energy and have the correct number of common quantum-mechanical features. Its energetic levels are at an identical distance from each other and take multiple values 0, $h\omega_0$, $2h\omega_0$, $3h\omega_0$, etc. The last level is called zero, since below it the energetic states in a macromolecule do not exist. The upper boundary can take any value - from one to 100, 1000 and more energetic levels. Naturally, the higher the number of levels in the DNA-wave, the larger the cell's energetic potential. This form of cell energy we call biological also. During ontogenesis the amount of the biological energy varies from a maximum (energetic level of the sporophyte initial of a species) to a zero level (energetic level of a natural death of a plant) (Kuznetsova, 2003) (Figure 1-a). But during each concrete moment, it can take only one of its discrete values.

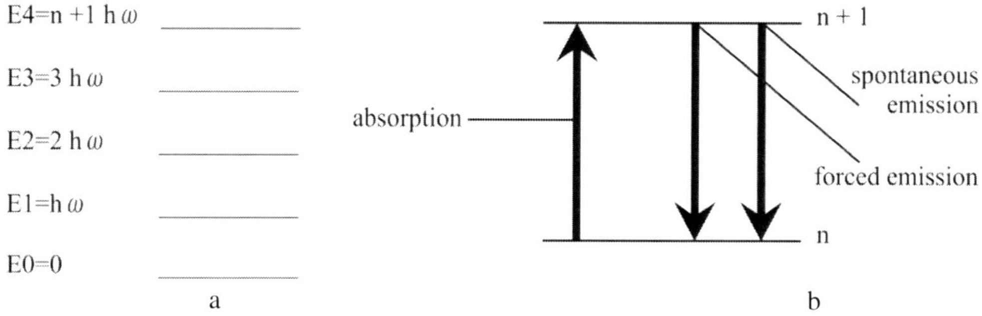

Figure 1. Energy levels of biological harmonic oscillator (a) and transitions between its two energy levels (b).

It is well known that the wave phenomena in DNA are not registered. But this does not deny the hypothesis about DNA as a particle-wave. This can be confirmed, first, by the imperfection of the existing methods and, second, by the fact that the DNA-wave exists only in living cells, whereas the majority of biological methods are based on cell destruction.

Physical and Biological Harmonic Oscillators: The Common and Distinctive Features

The presence of discrete energetic levels is a common characteristic of all harmonic oscillators. The transition of the system to higher or lower levels of energy is connected with the processes of absorption and radiation of free energy quanta. The fundamental feature of harmonic oscillators is that they, possessing a specific energy, do not conserve this energy. Therefore they periodically radiate it in portions, or quanta. In the course of energy loss, the harmonic oscillator is forced to descend from level to level, and the whole system, according to the 2nd law of thermodynamics, tends to balance. This is the reason why the echo in the forest quiets down, why the swing stops, and why the plant's life ends in death. Further analysis of cells and plants has shown that their properties completely coincide with quantum harmonic oscillator characteristics.

The principal distinction between biological and physical harmonic oscillators is as follows: In physical systems, the processes of absorption and radiation of energy are equally probable. For example, it is possible to set the swing in motion again. In biological systems they are separated. A consequence of a separation of the processes into two energetic flows, opposite in direction – ascending and descending, is the two-period life of multicellular organisms – sexual reproduction and ontogenesis, which are equally significant for plants (Figure 2). The reproductive period is connected with energy absorption and the transition of the DNA-wave to higher levels of energy. This process is genetically determined and demands considerable energy expenditure.

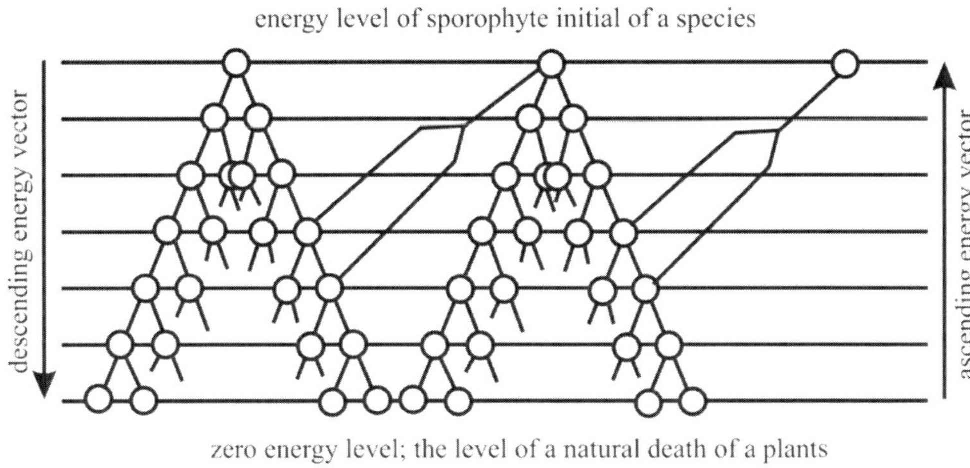

Figure 2. The scheme of life is continuous during the alteration of a generation (ontogenesis and sexual reproduction periods).

The fixed level of biological energy is achieved on the basis of an energetic line of the reproductive cycle of the species (Kuznetsova, 1998, 2002, 2003). The reverse process – energy radiation and transition of the DNA-wave to lower energetic levels takes place in ontogenesis.

For the most part somatic cells this way are unidirectional. The biological energy decrease will continue until the DNA-wave reaches the zero level, which means the end of a cell's individual life on the basis of one DNA-particle-wave.

The second difference, due to the fact that whereas in physical systems the free energy released by radiation disperses, in biological systems the energy of the DNA-wave is absorbed and spent on useful work in a cell. But since the efficiency cannot be equal to 100%, part of the inner energy is lost. Its traces were registered as a thermal flare in the interphase period (Zholkevich et al., 1973). This loss of heat was so great that the cell's energy balance became negative, which characterizes the amount of energy coming from the DNA-wave into the cell.

Hence, on the assumption of the common quantum-mechanical features of oscillating systems and cell biology data, the cell should be a quantum-mechanical system.

Operational Modes of a Biological Harmonic Oscillator in the Cells and Plants

In our model a cell is a quantum-mechanical system. The operation of all quantum-mechanical systems is a function of such physical factors, as light, pressure, temperature and volume (Feynman et al., 1863). The volume of the nucleus and cytoplasm in the cell are determined by a nucleocytoplasmic ratio. Hence, for the analysis, there remain only three variables.

Light

Einstein proved that the probability of an atom's radiation depends on the intensity of the light [cited from Feynman et al., 1963, vol.1]. He described three types of processes: 1) absorption, 2) energy radiation proportional to the intensity of the light around and 3) spontaneous radiation, which are not dependent on light (Figure 1-b). If we extend this knowledge to the operation of a biological harmonic oscillator, then it becomes clear, why light is represented as a factor in the organization of the life of plants. Thus, the interaction of light with the DNA-wave explains the nature of such biological phenomena as photoperiodism, rhythms of mitotic activity, etc.

Temperature

It is established that the radiation of a quantum harmonic oscillator fades away at low temperatures, sharply drops at high temperatures, and in an average (effective) region is a function of kT. Since a cell is a quantum-mechanical system, the DNA-wave radiation can take place only in the effective temperature region (0-40^{0}C). In this region the speed of biochemical processes grows and the duration of a cell cycle decreases in proportion to the temperature (Mazia, 1961; Stryer, 1981; Grift et al., 2002). Beyond its boundaries, where the DNA-wave as a physical body is nonfunctional, the cells are non-viable. The latter explains the existence of rest in plants in winter and in tropical and subtropical regions during the drought season.

Pressure

There is a similar relation between the operational modes of a biological harmonic oscillator and the intracellular pressure, since the temperature and pressure by the gas law are directly related. But the plant cell imposes additional, stricter requirements on this parameter. The analysis of a generative cells formation showed, that pressure is a major factor, by means of whose forces the energetic state of the DNA-wave during sexual reproduction period grows. Therefore, the physical guarantee of unidirectionality and irreversibility of ontogenesis can be provided only by the conditions of physiological homeostasis. The absence of homeostasis gives the chaos caused by frequent changes of development programs of vegetative and generative cells in one plant. For this reason all cells of plants characterized the constancy of osmotic pressure, pH medium, etc., which are sustained at a same species-specific level throughout its life.

Sources of Inner Energy of a Cell

According to the existing model (corpuscular), the cell is an open system. It does not tend to equilibrium; therefore it should constantly absorb energy from outside. Energy is received from two outer sources – photosynthesis and respiration.

Our model (corpuscular-wave) is based on the concept of the dual nature of DNA and on the notion of a cell as a quantum-mechanical system, the open and closed system, the equilibrium system obeying the fundamental laws of physics, including the laws of thermodynamics. The cell cycle energetics is provided by 4 independent sources: inner – the energy 1) of the DNA-wave and 2) metaphase chromosomes; and outer – the energy 3) of photosynthesis and 4) respiration (Figure 3). In the literature the outer sources are well described, therefore, we will dwell only on the principle work and physical nature of the inner ones.

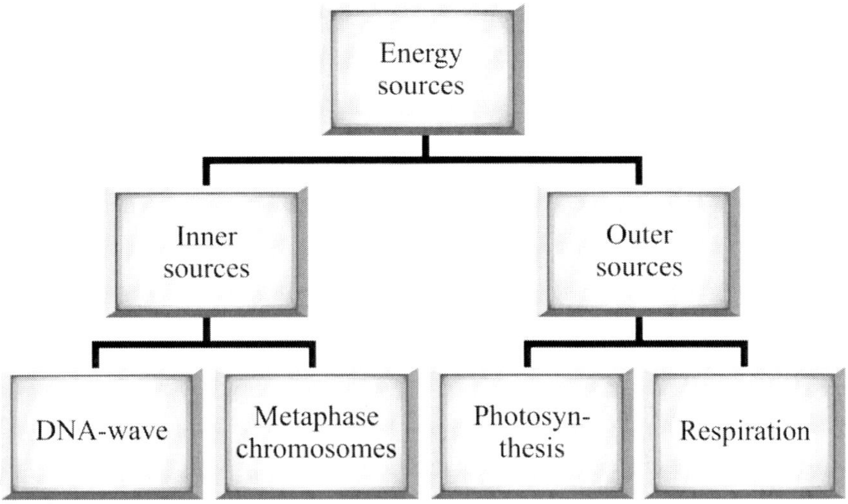

Figure 3. The internal and external sources of free energy in a plant cell.

Biological Harmonic Oscillator or DNA-Wave Energy

Energy as the physical phenomenon can exist in different forms: kinetic, chemical, nuclear, etc. The biological energy by its physical nature represents the energy of a quantum harmonic oscillator. A nucleus (in a prokaryote the whole cell) together with the oscillating DNA function in the cell are a united oscillating system (biological harmonic oscillator), as an inner source of free energy (biological energy). The cells and multicellular organisms have learned to use the mechanism of the two-sided transition of potential energy into free energy within the operational framework of the biological harmonic oscillator, as the physical body. Its work can be compared to an "energetic reservoir", which is filled with energy to the species-specific level in the reproductive development of an individual, and is released from it in portions during its ontogenesis. This source emerged with the appearance of life on the Earth and meets in all types of cells: vegetative and generative, proliferous and differentiated, prokaryotic and eukaryotic, of a plant and animal origin.

The periodic radiation of the DNA-wave divides (splits) the living matter into cells. In a cell cycle the DNA-wave radiation apparently falls on the end of the G_1 period. It is experimentally established that during interphase, a powerful thermal flare takes place (Zholkevich et al., 1973) and on S period the first maximum of the ATP pool level can be found (Mazia, 1961; Strochkova and Kazan'ev, 1975). Portions of free energy periodically entering into a proliferous cell are spent on DNA doubling and mitotic center formation, thus initiating one cell division after another. And this will go on until the biological energy bulk in the DNA-wave comes to an end. Hence, the presence or absence of biological energy is a feature distinguishing a living cell from a dead one.

The appearance of a biological harmonic oscillator, as an additional source of free energy has given the animate matter the advantage over the inanimate matter, and its continued development, but in a new form, as living matter. During evolution the DNA-particle and DNA-wave regularly varies. The DNA-particle elongated, divided and structurally improved, which lead to the growth of the genome volume (Swanson and Webster, 1977; Carroll, 2001; Alberts et al., 2002; etc.). The DNA-wave mastered level after level, creating thus a great number of more complicated and at times cardinally reconstructed matrices of reproductive cycles of the species of a different typical basis (gymnosperms, angiosperms), each of which is realized in ontogenesis in the species-specific plants classified by the taxons of different categories (genera, classes, genus, etc.).

Metaphase Chromosomes Energy

The second inner source of free energy is the energy of metaphase chromosomes. This source emerged with the appearance of a nucleus and mitosis. It is present only in proliferous eukaryotic cells of plants and animals. The analysis of a biological phenomenon – a metaphase chromosome - showed that during the coiling of chromosomes we deal with a physical phenomenon – the accumulation of elastic energy inside the super-compressed spring of DNA. By Hooke's law, the amount of the energy reserved in it is determined by the formula $U = kx^2/2$, where x is a degree of the spring's compression. Imagine this amount of energy, which concentrates inside the metaphase chromosomes, if at coiling the super-coils of the 4th order are created, and the chromosomes are shortened by 10^2-10^4 times (Stebbins, 1971; Stack and Anderson, 2001; Chentsov, 2004). By the cell scale this is, certainly, a great amount.

The reverse process – energy release at despiralization of chromosomes - is very rapid. This means an "energy explosion" inside the closed system –dividing a cell in telophase. On the basis of experimental data we will track the transformation of energy from the second inner source into other types of energy. One part of it is spent on the energetic securing of a transcription process and is reserved in the chemical bonds of various types of RNA. It is known that during late telophase the RNA biosynthesis proceeds intensively on euchromatin (i.e., the most uncoiling) regions of the DNA strand (Swanson and Webster, 1980; Alberts et al., 2002, etc.). The second part is transformed into the kinetic energy of cell molecules. At the expense of this energy the work directed at the increase of the volume of a cell, which follows its division, is done (period G_1 of the interphase) (Mazia 1961; Anderson et al. 1969). The third part is spent on the creation of an electromagnetic field; this causes the nucleic envelope and DNA-wave to be restored. The fourth part is used on the restoration of ATP. During the cell cycle the ATP pool level decreases during division and its second boundary falls on a period of cytokinesis (Mazia, 1961; Chin and Bernstein, 1969; Strochkova and Kazan'ev, 1975). The last part of the energy is spent on the excitation of biomolecules. When they return to the initial state, it escapes as fluorescent radiation. This radiation was revealed in the dividing cells by Gurvich (1932), who called it the mitogenetic radiation.

Corpuscular and Wave Properties of the Living Matter at Different Levels of Its Organization

The analysis of the experimental material showed that the data is completely in keeping with the corpuscular-wave model. The dual nature manifests itself both in the structure of biological systems, and in the development of biological processes. In support of this statement a short twofold characteristic of the living matter according to the hierarchical levels of its organization is given below.

- *On the molecular level.* a) In the DNA structure: DNA is a particle (a molecule of deoxyribonucleic acid) and a wave (with amplitude and oscillation frequency). b) In the DNA replication mechanism: the DNA-particle serves as a matrix for the replication process, but the accuracy of reading the genetic information is provided by the DNA-wave.
- *On the submolecular level.* a) In the chromosome structure: the chromosome is a particle (length, arrangement of nuclear organoids) and a wave (a set of standing waves). b) In the mechanism of chromosome reproduction: the chromosomes-particles double in the interphase (S period) and are separated from each other in anaphase. The chromosome-wave is present in interphase (cell - open system), but is absent during mitosis (cell - closed system).
- *On the cellular level.* a) In a cell structure: a cell is a particle (a structure with a nucleus, a cell envelope, cytoplasm and organelles) and a wave (a quantum-mechanical system, a source of biological energy and a biofield). b) In the mechanism of cell division: mitosis ends with the formation of the daughter cells identical to the mother cells in DNA-particles (hereditary basis of a cell), but differing from it by one energy level of DNA-wave (energetic basis of a cell).
- *On the organism level.* a) In the structure of a reproductive cycle and genotype: the reproductive cycle consists of two lines for the independent formation of the DNA-particles (hereditary basis) and the DNA-waves (energetic basis) in a genotype of the initial sporophyte. b) In the life of plants: a cell (particle) is an elementary unit of a multicellular organism, but the development of cells in space and time is a wave, extending periodically during the course of ontogenesis, and narrowing periodically during the course of sexual reproduction, but always preserving the corpuscular-wave specificity of a species.
- *On the species level.* a) In the structure of a species: a species is a particle (individuals with the species-specific DNA-particles) and a wave (individuals with the species-specific set of energetic states of the DNA-waves). b) In the mechanism of a stabilizing selection: the pool of the genotypes (particles) of pollen, egg cells, embryos, plants is an object of selection, but their motion in the life cycle of generations is a wave periodically extending by means of superfluous breeding , increasing variability (genotypic, phenotypic) and periodically narrowing by means of selective survival, but always preserving the corpuscular-wave norm of a species.
- *On the biospheric level.* a) In the structure of living matter: the living matter is a particle (a set of proteins, exchange of substances) and a wave (split into cells, divided into the reproduction and ontogenesis periods). b) In the evolution

mechanism: an object of evolutionary change is a species (particle), but the process of its evolution is a wave, which periodically extends due to the appearance of new species on the basis of a reproductive cycle matrix of one type, and periodically narrows when creating a matrix of the next type, but always directed at the increase of the biological energy of species.

References

Alberts, B., Johnson, A., Lewis, J., Raff, M., Roberts, K. and Walter, P. (2002). *Molecular biology of the cell*, 4th edn. Chaffey, N (Ed.), Garland Science, New York.

Allen, J.P. (2008). *Biophysical chemistry*. Wiley-Blackwell Publ.

Anderson, E.C., Bell, G., Petersen, D. and Torbey, R. (1969). Cell growth and division. 4. Determination of volume growth rate and division probability. *Biophys.*, 9(2), 246-263.

Bauer, E.S. (1935). *Theoretical Biology*. Moscow, Leningrad.

Brachet, J. (1957). *Biochemical cytology*. Academic Press, New York.

Cantor, C.R. and Schimmel, P.P., 1980. *Biophysical chemistry*. Freeman, W.H. (Ed.), San Francisco.

Carroll, S.B. (2001). Chance and necessity: the evolution of morphological complexity and diversity. *Nature*, 409(6823), 1102-1109.

Chentsov, Yu.S. (2004). *Introduction to cell biology*. Akademkniga, Moscow.

Chin, B. and Bernstein, I.A. (1969) Adenosine triphosphate and synchronous mitosis in *Physarum polycephanum*. *Cell Biol.*, 96(2), 330-337.

Cremer, T., Kreth, G. and Koester, H. (2000). Chromosome territories, interchromatin domain compartment and nuclear matrix: an integrated view of the functional nuclear architecture. *Crit. Rev. Eukaryot. Gene Expr.*, 10, 179-212.

Feynman, R.P., Leighton, R.B. and Sands, M. (1963). *The Feynman Lectures of Physics*. vol. 1-3. Wesley, A. (Ed.), Readigg, Massachusetts, Palo, Alto, London.

Grift, I.G., Ivanov, V.B. and Machs, E.M. (2002). Cell cycle and its parameters in flowiring plants. *Tsitology*, 44(10), 936-980.

Gurvich, A.G. (1932). *Mitogenetic radiation*. Medgiz, Moscow.

Gurvich, A.G. (1944). *Theory of biological field*. Sovetskay nauka, Moscow.

Gurvich, A.G. (1977). *Selected works*. Moscow, Meditsina.

Hughes, A.F.W. (1959). *A history of cytology*. Schuman, A. (Ed.), London.

Kuznetsova, N.F. (1998). Activation in the course of reproductive development in female generative sphere of *Pinus sylvesrtis L. Tsytology*. 40(10), 849-854.

Kuznetsova, N.F. (2002). The reproductive development of initial sporophyte of *Pinus sylvesrtis L. Tsitology*, 44(2), 159-165.

Kuznetsova, N.F. (2003). The formation of the sporophyte initial of *Pinus sylvestris L.* as deduced from the deciphering of the reproduction cycle of the pine. In: *Mater. XI s'ezda RBO*, vol. 2. Barnaul, 18-22 August 2003, 149-150.

Lehninger, A.L. (1965). *Bioenergetics*. Benjamin, W.A. (Ed.), New York.

Lehninger, A.L. (1985). *The principles of biochemistry*. V. 1-3. Moscow, Mir.

Mazia, D. (1961). *Mitosis and the physiology of cell division*. In: Brachet, J. and Mirsky, A. (Eds.), Academic Press, New York, London.

Prescott, D.M. (1970). The structure and replication of eukaryotic chromosomes. *Adv. Cell Biol.* 1, 57-117.

Rubin, A.B. (1999). *Biophysics*. V. 1-2. Nauka, Moscow.

Stack, S.M. and Anderson, L.K. (2001). A model for chromosome structure during the mitotic and meiotic cell cycles. *Chromosome Res.*, 9, 175-198.

Stebbins, G.L. (1971). *Chromosome evolution of higher plants*. Arnold, E. (Ed.), London.

Strochkova, L.S. and Kazan'ev, V.V. (1975). Cytochemical peculiarities of a cell cycle. *Cytology. vol. 2.* pp. 9-66, VINITI AN SSSR, Moscow.

Stryer, L. (1981). *Biochemistry*, 2nd edn. Freeman, W.H. (Ed.), San Francisco.

Swanson, C.P. and Webster, P.L. (1977). *The cell*, 4th edn. Prentice-Hall Inc., Englewood, Cliffs, Jersey.

Volkenshtein, M.V. (1965). *Molecules and life. Introductions to a molecular biophysics*. Nauka, Moscow.

Volkenshtein, M.V. (1983). *Theoretical biophysics*. Nauka, Moscow.

Zholkevich, V.N., Gurvich, A.L., Borisova, T.L., Livanova, T.N. and Shlyakhtina L.S. (1973). On energy discharge prior to cell division. *DAN SSSR*, 209(4), 961-965.

Chapter II

Cell Cycle

Abstract

In this chapter the corpuscular-wave model of the cell cycle of proliferous eukaryotic cell of plant is discussed. The cell cycle has been divided into two periods, four stages and 12 parts. The alterations of DNA-particle, DNA-wave, environment, main structural components and functions are analyzed separately for each stage of the cell cycle. The development of the cell from birth to death is represented as a united biophysical and biochemical process, during which it is independently and orderly changed into a DNA-particle and DNA-wave.

The transition of a DNA-wave to a lower energy level and the quantity of free energy entering into a cell initiates the process of the formation of new cells (replication of DNA, origination of centrioles, condensation of chromosomes, etc.). The first stage of their formation (a stage of latent development) occurs inside the mother cell, which at this time is an open system and has the ultrastructure and metabolism of an interphase cell. After replication the DNA is doubled, and the cell genome then consists of 2 DNA-particles - 1 DNA-wave. The next stage (mitotic development of cell) begins with the destruction of the DNA-wave, biofield, nucleus, a splitting of chromosomes on chromatids, a transition of cell from an open system to a closed system and a growth of entropy. The genetic material is represented by 2 DNA-particles - 0 DNA-waves. At this time its development goes as the physical (corpuscular) process in the condition of closed system. As a consequence, the chromatids of metaphase chromosomes at the end of each metaphase have the equal quantity of elastic energy and the identical DNA-particles. In the stage of cell birth the chromosomes are separated and diverge to the places of the formation of future cells. The chromosomes begins to untwist, the liberated free energy enters into a telophase cell and increases its energy. By means of the forces of the electromagnetic pole a nucleus envelope is reestablished and it is arranged like biomolecules in the intracellular volume. The mitosis ends with the division of the cell body. The DNA molecule exists as a 1 DNA particle - 0 DNA-wave. The cell is a closed system. The ontogenesis is the last and the longest stage of cell cycle. It begins from the restoration of DNA-wave, biofield, centrifugal and centripetal energy streams, cell growth, metabolism and ultrastructure of interphase cell. During interphase the genetic material can be in two forms: 1 DNA-particle - 1 DNA-wave and 2 DNA-particles - 1 DNA-wave. Thus, the cell cycle of eukaryotic cell is organized so that the each successive cell generation receives from the previous generation the identical DNA-particle and the DNA-wave with the energy lower on the one level.

The cell cycle, as an independent biological phenomenon, is characterized by a strict sequence of processes, continuous chemical activity and ordered energy streams. According to current views, it divides on 4 periods: pre-synthetic (G_1 period); synthetic (S period); pre-mitotic (G_2 period) and mitosis (M). The duration of mitosis is only a tenth of the whole cell cycle (Epifanova, 2003). A good amount of experimental work has been done in this area and a huge amount of material on different plant species has been accumulated. Nevertheless there is no explanation of the physical nature of changes during mitosis and the cell cycle of proliferate cells. In the present chapter we have tried to find in the biological facts their physical essence, and then construct a corpuscular-wave model of a life cycle of plant cell, whereby we can determine the changes in the DNA-particle, DNA-wave, energy streams, intercellular processes and events.

The principle difference of our model from the customary ones is due to its basis in the new theoretical approach of cell organization and has put in priori – a corpuscular-wave nature of DNA and the quantum-mechanical nature of a cell. The analysis of intracellular processes and events of development of a proliferous cell is carried out by the fundamental laws of physics and experimental data of cell biology. Moreover, having analyzed this large amount of experimental material, we have not found any fact that would not keep within corpuscular-wave cell cycle model.

In our model, the initiation of the development of new cell generations is connected with the wave properties of DNA. It is a transition of the DNA-wave to the lower energy level and an entering into a cell of the quantum (portion) of free energy. Such beginning of a cycle has the following experimental and theoretical confirmations. 1. It is established, that the initiation of DNA replication occurs once in a cell cycle (Lark and Lark, 1979). Suppression of protein, RNA synthesis and an increase in the weight of a cell does not have any influence on the initiation mechanisms. The DNA-wave radiation is purely a physical phenomenon, which is characterized by periodicity and spontaneity. Therefore periodically radiating DNA-waves can be considered as the starting mechanism and an energy source for the beginning of the proliferation processes. 2. During interphase a powerful thermal flash at the synchronized cultures *Panax ginseng* and *Torula utilis* is registered (Zholkevich et al., 1973). A cell at this time is an open system. Therefore the work of biological harmonious oscillator (a nucleus together with oscillating DNA), as an internal energy source, can be revealed and estimated by means of a quantity of thermal losses. 3. It is proved, that DNA synthesis and mitosis are the most energetically dependent processes of a cell cycle (Webster and Vant'Hof, 1969). The energetic sources of both processes in corpuscular biology are not known. The biological harmonious oscillator is capable of providing the reaction of DNA doubling by free energy, if we take into account the amount of heat loss determined by Zholkevich et al. (1973). 4. A DNA-wave is a standing wave. During radiation the energy goes into a cell by two streams. Therefore two independent processes of the formation of two cells are initiated simultaneously.

A life cycle of a proliferous plant cell on the 12 stages, 4 quarters and 2 periods are divided. The short description of stages is given in table 1. As is known, a number of intracellular mechanisms (doubling – division of DNA-particle; formation – destruction of mitotic apparatus; condensation – decondensation of chromosomes; etc.) have a cyclic characteristic. During the evolutionary period of cell cycle there is an unfolding of these processes, during involutionary periods, on the contrary, is a folding in of these processes... The genetic material in the system of a DNA-particle and a DNA-wave and the energetics of cell are changed once in a cycle quarter. Therefore we will describe a cell cycle by the stages (or cycle quarters), which have been named as: 1) latent development; 2) mitotic development; 3) cell birth; 4) ontogenesis.

Table 1. The main intracellular events during life cycle of eukaryotic proliferous cell of plants

1st period of cell cycle – evolutionary development
1st cycle quarter – the stage of latent development

N	Leading event	Cell cycle stages	System type	List of intracellular events
1.	Initiation division	Interphase, G_1 period (end)	Open	Irradiation of DNA-wave; formation of two (+)-changed centrioles and a number of (-)-changed active DNA-polymerases; cell divide on nucleus, cytoplasm and organelles; continue the processes of respiration, photosynthesis, synthetic processes
2.	Replication DNA	S period	Open	DNA-article doubling by complementary principle
3.	Chromosome coiling	G_2 period	Open	Beginning of 2nd source of internal energy creation by means internal coiling of chromosomes

2nd cycle quarter – the stage of mitotic development

4.	DNA-wave destruction	Early prophase	Close	Destruction of DNA-wave, biofield, nucleus, nucleolus; increase of entropy; respiration, photosynthesis, synthetic processes ceases; centrioles diverge; chromosomes split on chromatid
5.	Chromosome condensation	Late prophase	Close	Creation of 2nd source of internal energy by means external coiling of chromosomes; chromosome motion to equatorial region of cell
6.	2nd source of internal energy creation	Metaphase	Close	Formation of equatorial plate; equal distribution between chromatid hereditary material and energy for future cells

2nd period of cell cycle - involutionary development
3rd cycle quarter – the stage of cell berth

7.	Chromosome movement to centrioles	Anaphase	Close	Division of centromere; separation of chromatid and chromosome motion to centrioles; respiration, photosynthesis, synthetic processes absence
8.	Chromosome decondensation	Early telophase	Close	Neutralization of changes; destruction of centrioles, a structure with a charge transfer, electric field; beginning of chromosome decondensation; appearance of electromagnetic field; entering of energy from metaphase chromosomes into cell

(Continued)

9.	Reconstruction of nucleus	Late telophase	Close	Reconstruction of nucleus envelope, ordering of molecular compound; decrease of entropy; realization of cell cytokinesis; initiation of centrifugal and centripetal energy streams

4th cycle quarter – the stage of ontogenesis

10.	Ontogenesis beginning	Interphase beginning	Open	Reconstruction of DNA-wave, biofield, interphase cell ultrastructure by its DNA-particle and DNA-wave; beginning of respiration, photosynthesis, different synthetic processes, etc.
11.	Ontogenesis	G_1, S, G_2 periods	Open	Doubling of cell volume, quantity of proteins, lipids and other biomolecules; support of centrifugal and centripetal energy streams on optimal level
12.	Ontogenesis end	Interphase end	Open	Progressive decrease of centrifugal and centripetal energy streams level

1st Cycle Quarter– The Stage of Latent Development (2 DNA-Particles – 1 DNA-Wave)

According to our notions a free energy entering into proliferous interphase cells from DNA-wave (1st source of inner energy) somewhere at the end of G_1 period utilizes by the following channels (Figure 4). The main part of energy is spent on a DNA replication. The second part is expended on a centrioles formation. During chromosome coiling this energy is transformed in the energy of 2nd internal source (metaphase chromosomes). The third part of the energy is turned into the kinetic energy of cytoplasm molecules, causing cell growth during G_2-period. The experimental data testified that the cell growth is accelerated twice – during the G_1 period and during the S-G_2 periods (Terasima and Tolmach, 1963; Alov, 1975). The last part of internal energy is passed on for the reconstruction of ATP from ADP. Increased ATP content in cells during S period is confirmed as well (Chapman et al., 1971; Murphree et al., 1974). During interphase the energy of high-energy bonds will be consumed in order to provide an energetic balance of the cell. The rest of the energy is lost in the form of heat (Zholkevich et al., 1973).

A centriole is one of the most significant organoids for cell division (Mazia, 1966; Pawelets et al., 1984; Alberts et al., 2002; Chentsov, 2004). The divergence of centrioles creates in a dividing cell the polarity that is required for mitosis realization. However they still have not found it in a plant cell. The data about a time of centriole origin, chemical compound, localization and function during cell cycle are very contradictory. We have all the foundations for an assumption that this pair structure arises in the time of DNA-wave radiation, and its main chemical component is the actin. This protein does not meet in prokaryotic cells, where mitosis is absent. But the actin is always contained in eukaryotic cells (Watson, 1965), where it composes more than 1/5 of all the proteins of a cell (Actin: A dynamic, 2000). The actin reveals itself among the proteins of the mitotic apparatus (Mazia, 1961; Gawadi, 1971; Swanson and Webster, 1977; Ishizaki et al., 2001, etc.). In a nucleus and cytoplasm it forms a thick and labile network of microfibrils (Lark and Lark, 1979; Cell cycle control, 1995; Vasilicv, 1996; Tichonov, 1999; Welch, 1999; Alberts et al., 2002; van Driel, 2003; Kiseleva and Allen, 2004; Pederson and Aebi, 2005; etc.). All actin-related proteins (Arp) are a very conservative. In all investigated organisms they consist almost exclusively from an identical set of amino acids and is characterized by G – F actin transitions. The mechanism of its polymerization-depolymerization is reversible, so the system microfilaments are in a condition of dynamic instability. About half of the actin molecules are in the form of monomers, and half is moved into the structure of microfilaments (diameter≈ 6-7nm). The mechanism of the rapid local polymerization-and depolymerization of actin is observed in many processes of cytoskeleton dynamics (muscle contraction, cell motion, etc.).

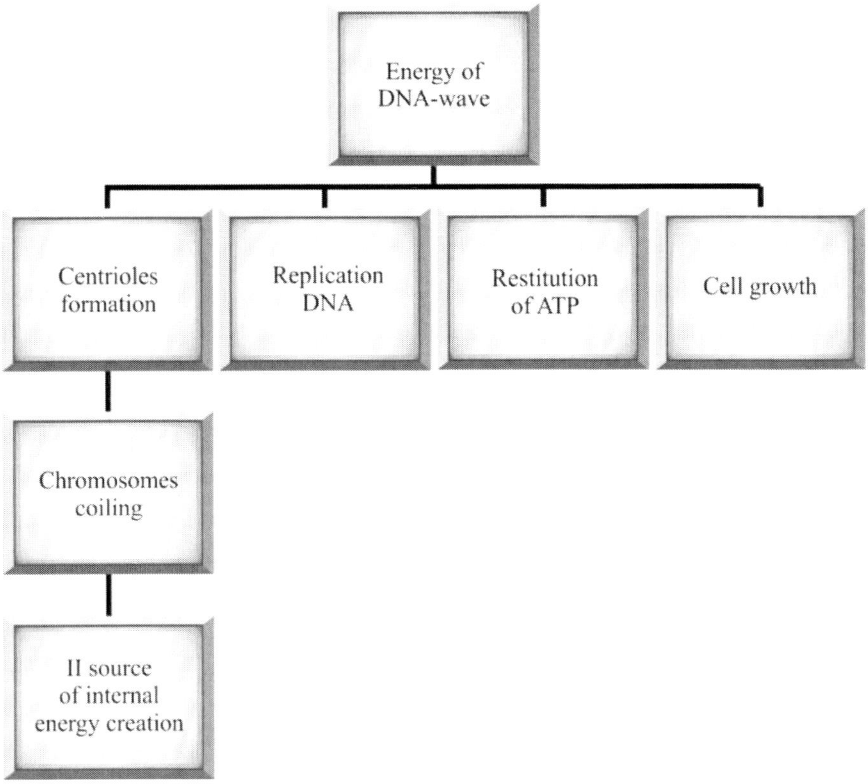

Figure 4. The scheme of utilization of energy from a DNA-wave.

Based on the data listed above, we suggest the following mechanism of centrioles formation. The origination of centrioles is initiated by the energy of a DNA-wave, which enters into the nucleoplasm in the end of the G_1 period. The actin microfilaments in the zones of direct action of the two energy streams are quickly polymerized. Simultaneously these energy streams beat out electrons from actin molecules and give them some energy. The raised electrons jump from the actin on a DNA-polymerase and transfer this enzyme in an active condition. The actin molecule then becomes positively charged, while the active molecules of the DNA-polymerase are negatively charged.

An inactive enzyme of the DNA-polymerase at this time is in the internal areas of the nucleus (Paponov, 1987). A presence or an absence in a nucleus of active DNA-polymerase is the main factor, limiting the process of replication (Friedman, 1970; Harland et al., 1973; Kornberg, 1980; Stryer, 1981; Karp, 1996; etc.). It is established, that a DNA-polymerase activity always increases with the initiation of DNA synthesis. A DNA-polymerase, having received energy and a charge, will start to function as an intermolecular transfer of energy (Riegel, 1967).

Thus, during DNA-wave radiation three main events, which are required for mitosis realization, have occurred: 1) the two positively charged threads of a polymeric actin were formed; 2) a large number of negatively charged molecules of active DNA-polymerase has appeared; 3) a chemical structure with a charge transfer was created. This structural chemical compound with a charge transfer will act from the S period interphase to the telophase of mitosis. All this time the exact sequence of electron transport is set.

A DNA molecule, by its physical properties, is thin, hard and a double spiral (Volkenshtein, 1965, 1983; Rubin, 1999, Allen, 2008; etc.). It is a neutral molecule as a whole (Volkenshtein, 1983; Cantor and Schimmel, 1980). However the nucleotides have a dipolar structure. The negative charges are concentrated in the region of phosphate groups, and the positive charges are in the region of bases.

The negatively charged active DNA-polymerase after activation "sits down" on one of DNA threads in the region of bases where there is a zone of positive charges and starts to move passively with one of its waves. The strong field of enzymes breaks the electric balance in the place of a molecules union. As the association is a negative charge, the negatively charged phosphate groups of two DNA threads push off from it. The increased asymmetry leads to local destruction of the hydrogenous connections between threads. In this area the threads are then straightened, untwine, move away from each other and form the loop, which will function as a DNA-replication fork (Drake et al., 1969).

The molecules of an active DNA-polymerase, which are moved together with a DNA-wave, automatically untwine the DNA threads in the direction of the movement. The nucleotides on a different wave at this time pass by in the opposite direction (Figure 5). The enzyme, expending energy, is the catalyst, the reaction of the complementary joining of nucleotides.

A DNA-wave is the standing wave. During radiation the energy goes into a nucleus by two streams and therefore the two blocks of DNA-polymerase molecules are simultaneously activated. Having received a charge, they (as same charged particles) are pushed from each other and are distributed along the macromolecule. Then the molecules of enzyme of two blocks are attracted to the contrary waves and to the different threads of DNA in the region of the bases and are moved in opposite directions. After that a DNA-polymerase in the 5' $\rightarrow$ 3' direction carries out the reaction of DNA biosynthesis into the regions of enzyme activity – the length of one replicon. The length of replicons may be different because during activation the enzymes receive a different quantity of energy.

After replication the inactive negatively charged molecules of DNA-polymerase continue to move with a DNA-wave. They pass in cytoplasm when the enzymes collide on a counter wave or with a nucleus envelope.

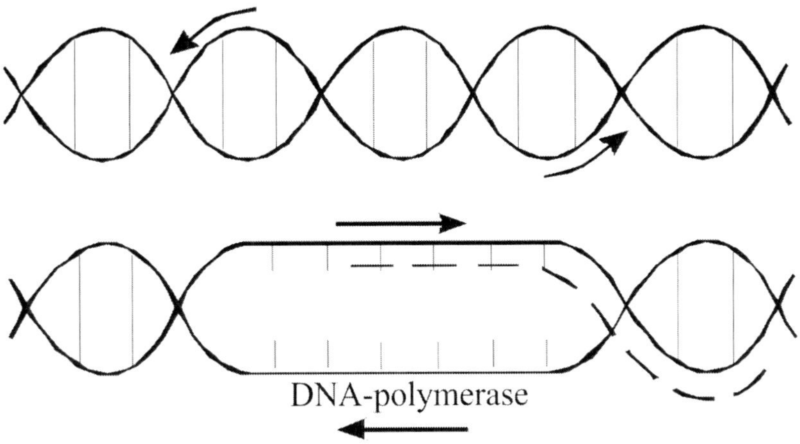

Figure 5. DNA replication scheme with participation of DNA-article and DNA-wave.

It is established, that the motion of a DNA-replication fork stops only after collision with a counter fork or on a chromosome end (Zhimulev, 2003). During the blow the electrons "jump over"from an enzyme to the chromosomes. Therefore a DNA is the neutral molecule from S period to telophase mitosis. During another part of the cell cycle (from telophase to G_1 period end) it is a negatively charged macromolecule.

The corpuscular-wave model of a replication mechanism is confirmed by numerous sources of experimental data. For example, the wave nature of DNA biosynthesis explains why the high exactness of genetic information reading, which is equal to $10^{-7} - 10^{-11}$ or less than one error for replication, is achieved (Drake et al., 1969; Tomilin, 1983). The reasons for an absence of replication at a fixed point on a DNA strand (Hand, 1978; Alberts et al., 2002), the wide range of differences of a replicon length – to 25 times are clear as well (Edenberg and Huberman, 1975). The asynchronism of a DNA biosynthesis reaction between the different chromosomes and within one chromosome is confirmed (Mueller and Kajiwara, 1966). The mechanism of the bilateral growth of a DNA chain ispolyreplicational, multistage and in a 5 ' $\rightarrow$ 3' direction is understood (Okazaki et al., 1969; Swanson and Webster, 1977; Ishizaki et al., 2001; Zhimulev, 2003, etc.). Why the DNA by one set of data is a neutral molecule (Haggis et al., 1965; Volkenshtein, 1965, 1983), and by another one – a negatively charged is explained (Mazia, 1957; Shakhbasov et al., 1986).

Today the existence of a turning isometry for organic compounds is completely proved (Volkenshtein, 1965, 1983). Therefore the DNA can be considered as the structure, which consists of the same links of turning isomers – the nucleotides. The nucleotides are also the elementary dipoles having the constant dipole moment at which the negative charge is displaced into the phosphate group, and positive. – to the base (Haggis et al1965;Volkenshtein, 1965, 1983). The nucleotides inside a DNA strand are packed one after another into piles, with a step of a spiral in 34Å and with a turn of 36^0 around a spiral axis. The piles, due to an electrochemical structure, present the greatest number of dipoles, which are positioned at an equal distance. The charges of the dipoles during formation of piles are orientated equally. It leads to an occurrence of polarizing charges along a chromosome arm – from a centromere to a telomere end. Most likely, these chains of positive and negative charges are characterized by the constant internal polarization, which are supported by their own field (Feynman et al., 1963). Thus, the electrochemical structure of normal DNA presents the 2 isomer piles, and a doubled DNA strand – the 4 isomer piles with, accordingly, 4 and 8 chains of positive and negative charges, polarized along the chromosome arm.

In a nucleus each chromosome has its own niche or its own "chromosomal territory" (Chentsov, 2004; Cremer et al., 2000; Monachova, 2004). The charges are oscillated together with the chromosomes-waves. Their motion generates the electromagnetic field, which divides a nucleus on two spaces – a near zone (changing as $1/r^2$, $1/r^3$) and a distant zone (accordingly, as 1/r). Therefore the oscillating chromosomes-waves it is a set of independent harmonic oscillators, working on the same distance from each other (Feynman et al., 1963). The maximum intensity of their summary electromagnetic field (biofield) is concentrated inside the small layer surrounding the chromosomes-waves. The centrioles (or the two positively charged molecules of polymeric actin, which during DNA-wave radiation were formed into a near zone of biofield), find themselves in strong electromagnetic field and its forces does not allow them to leave the given layer.

A strong electric field arises between centrioles and is concentrated inside a very narrow area. This field is external relative to the chromosomes passing through it. After replication the chromosomes represent 1 DNA-wave and 2 DNA-particles. Therefore one, two, three or four nucleotides or isomeric links can pass through a given electric field. And, according to the vague principle, each event has its own probability. During passage, through the electric field, 1-3 nucleotides in each of them are generated into an additional dipole moment, what is not altered is a conformation of the DNA molecule. The situation changes, when 4 isomeric links enter into the field (Figure 6). Under the influence of an external field an electronic polarization of the 1st and 4th nearest nucleotides to the centrioles (1 and 4) is increased. A polarization breaks the electrical balance in the 2nd and 3rd isomeric links and causes them to turn on 180^0, as the result of this action of attraction-repulsion forces between charges. After a turn the two interplane isomeric links (or 2nd and 3rd dipoles) are attracted to each other and their charges are neutralized. The strong external field induces a separation of charges inside the given united fragment. This causes a new isomer link. This link is complex in formation as three nucleotides in its structure are united. During this conformation change the macromolecule is slightly shortened, and its inner energy increases. Because the nucleotides disposition concerning centrioles and the succession of simple and complex isomers in a macromolecule are constantly changing, the uniform compression of DNA threads on all lengths is guaranteed. During the G_2 period a chromosome contraction by a one-sided nucleotide turns and a process of complex isomer formation is continued. As a result, a macromolecule is shortened gradually until it reaches critical size. The given process is not accompanied by an optical effect of condensation on chromosomes. Therefore, we named this type of chromosome spiralization as internal coiling.

Thus, the 1st cycle quarter, having begun from a DNA-wave radiation, a progressive chromosomes coiling by a one-sided nucleotide turn is finished (Table 1). The 1st part of the cell cycle we named the latent development, as the DNA biosynthesis and preparation to cell division go inside an interphase cell, in which the cytological appearance does not change. During this stage the cell is open system. It is characterized as a structured nucleus, cytoplasm and organelles. The base cell processes, respiration, photosynthesis, RNA, proteins synthesis etc. also continue (Mitchison, 1973; The cell cycle, 1973; Strochkova and Kazan'ev, 1975; Bell, 1980; Ivanov, 1990; Actin: A dynamic, 2000; Ishizaki et al., 2001; Alberts et al., 2002; Chentsov, 2004; etc.). DNA presents by 2 DNA-particles and 1 DNA-wave. However, increasing chromosome spiralization leads to the so-called irreversibility moment (Mazia, 1961) or to the critical moment (Zeuthen, 1964), when the cell division mechanism cannot be stopped.

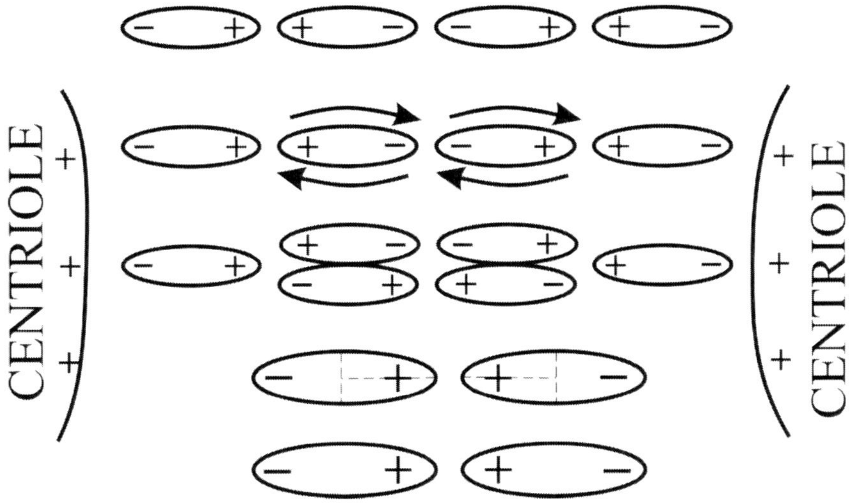

Figure 6. Chromosome coiling by internal revolving of DNA isomer links.

II Cycle Quarter - The Stage of Mitotic Development (2 DNA-Particles - 0 DNA-Wave)

The 2nd cycle quarter begins with the separating of chromosome-particles from a nucleus. Being shortened during coiling, they eventually become shorter than the length which the DNA-particle connected with a nucleus envelope in three points should be (Chentsov, 2004). At this moment the two chromosome-waves, not having reached a nucleus wall, disintegrate. The biofield, which is created by the oscillating DNA molecule, disappears also. The environment of a cell is fundamentally changed, when a DNA-wave and biofield are lost. The formation of future cells without a DNA-wave begins to carry out in the form of the one-plane (corpuscular) biophysical process, which was named as the stage of mitotic development. With the disappearance of the biofield forces, the two positively charged centrioles, which were kept into the inner area of the nucleus by biofield forces, mutuality repulse and fly away in opposite directions, giving rise to two cell centers. Therefore during two-, three- or four-polar mitosis the cell centers dispose in a cell geometrically correctly (Mazia, 1961; Sounson et al., 1969). A nucleic envelope starts to break up, because the intracellular disposition of phospholipid molecules is not fixed by a biofield force.

In prophase the doubled DNA-particle (condensed by internal coiling) represents the two piles of complex isomers or the two piles of dipoles in which charges are polarized along chromosome arms. Therefore chromosomes of such electrochemical construction have two vertical chains of negative charges outside of the piles, and two vertical chains of positive charges inside of the piles, which are balanced by means of hydrogen connections. That is why when centrioles or the two positively charged molecules of polymeric actin fly away, the chromosomes by external (negative) sides are pulled to the contrary cell centers and get torn on the chromatids up to the centromere regions. The chromosome splitting will occur by the hydrogen connections, or by a semi-conservative way of reproduction.

Between the centrioles there should be an electrical field, as between two charges having an identical quantity. In this case the power lines in the absence of charges between them look like as in figure 7a.

After splitting, a chromatid represents the single pile of dipoles with two vertical chains of polarized charges – a positive and a negative. The chromatid, which finds itself in the electrical field between two centrioles, reacts to it. They try to turn a negative side to the cell centers, but, by inertia having passed this position, are repulsed by the positive side, then again are attracted by the negative side. The two chromatids start in depend of rotation and a chromosome coiling renews. The in depend of rotation of the chromatids promotes their isolation from each other. As the coiling process occurs on the bases of complex isomers, the morphological effect of chromosome shortening is more obvious. The given type of chromosome coiling has been named by the external coiling.

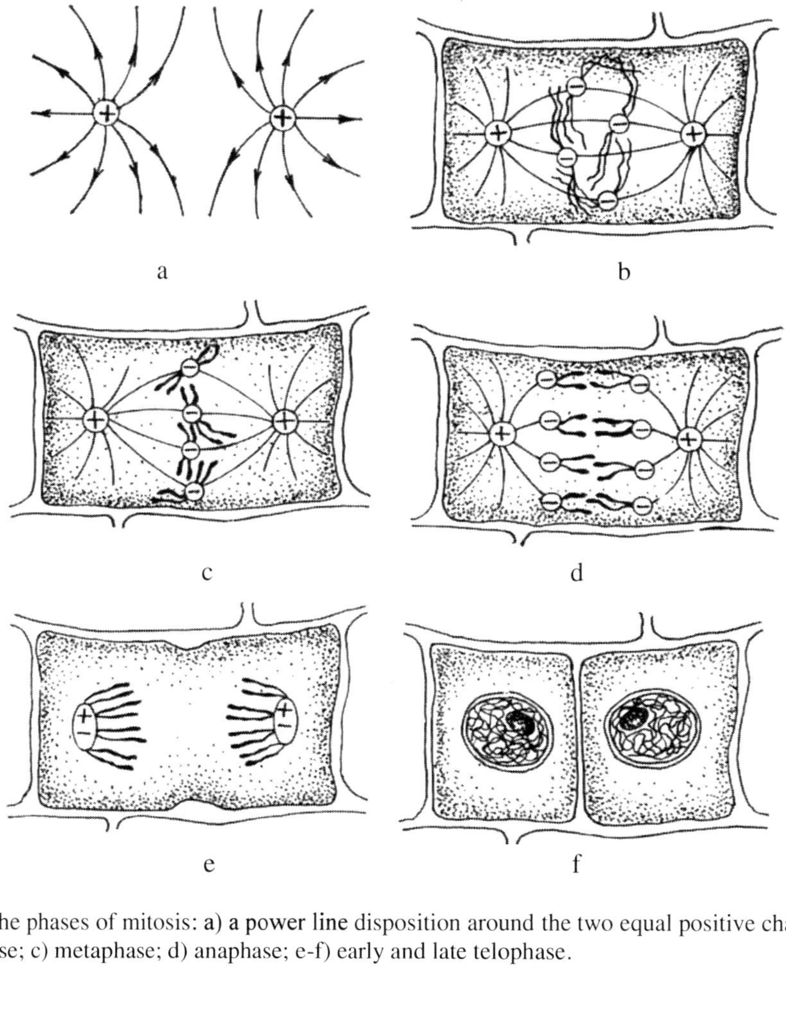

Figure 7. The phases of mitosis: a) a power line disposition around the two equal positive charges; b) late prophase; c) metaphase; d) anaphase; e-f) early and late telophase.

During the time of this coiling type at the turn of the complex isomers, their charges will be neutralized, and the DNA gradually loses a dipole construction. Therefore, the chromosome fragment without a centromere falls out of a metaphase plate and is not attracted to the cell centers (Sounson et al., 1969; Stebbins, 1971). The chromosomes during rotation are pushed out by the forces of an electrical field to the equatorial plate where its intensity is minimized (Figure 7b). They are separated from each other by the distance that is defined by the electrostatic forces of repulsion on negatively charged centromeres. During metaphase, chromosomes are constant in size, form and position on the equatorial plate. It is evidenced by a high coordination of the division processes.

During the physics position the metaphase chromosome is the equilibrium state between forces, acting on a chromosome from the electrical field, and the forces counteracting them from the contractive DNA. As is known, the elasticity laws apply to all physical bodies that have the ability to recover their size and shape after removal of the deforming forces. Chromosomes are one of them. Up to metaphase are the processes of chromosome coiling (condensation) and then uncoiling (decondensation) during telophase. Consequently, during the coiling, work was carried out that has been stored in the form of potential energy. By Hooke's law the quantity of elastic energy equal to: $E = ½ \kappa x^2$, where x – is the degree of the spring compression. After coiling the DNA length decreases 10^2 - 10^4 times (Sounson et al., 1969; Stebbins, 1971; Swanson and Webster, 1977; Chentsov, 2004; etc.). Therefore, in metaphase, chromosomes are concentrated into a great deal of elastic energy.

Potential energy is converted into kinetic energy during chromosome uncoiling and is used for the accomplishment of useful work. Therefore, metaphase chromosomes in the cell cycle are regarded as the second (2nd) in depend of source of internal energy, which provides the daughter generations of cells free energy after chromosome and cell division. It should be emphasized that the emergence of metaphase chromosomes as 2nd source of internal energy is a very important evolutionary step in the development of eukaryotic cell.

The electrons which have remained on chromosomes after replication, at coiling are passed on a centromere or on a spherical conductor. It is known, that power lines can begin and finish only on charges. As the chromosome arms are neutral, a power line, having begun on the positively charged centriole, should end on the negatively charged centromeres (Figure 7c), that coincides with a protein orientation in the microtubules of a spindle (Schmidt, 1939; Mazia, 1957; Swann, 1961; Alieva et al., 2008; etc.). In other parts of a cell, where there are no charges, the power lines go into infinity, being shown in the form of astral rays. The astral rays are usually found in animal cells. In plants, they are usually absent. Apparently, this is caused by the peculiarities of a plant cell matrix, which prevents the occurrence of an induction interaction between tubulin molecules. Therefore, at a chromosome gap, their fragments are capable of in depend of motion (Sounson et al., 1969; Alov, 1972). At the time of distribution, chromosomes in which centromeres are absent, the power lines connected to them on all lengths.

In cell biology, questions of the nature, energetics and mechanisms of the formation of an achromatic spindle are most discussed. It has been established, that spindle microtubules are not continuous (Aspiz, 1975). A calculation of *Lithodesmium undulatum* microtubules has shown that there are twice as many at the equatorial plate, than at the poles (Manton et al., 1969). There is data to support that the spindle is made up of two semi-spindles (Paweletz, 1968). Microtubules are characterized by a constant diameter (25 nanometers), double refraction and formation by tubulin monomers (Vasilyev, 1996; Moritz and Agard, 2001;

Alberts et al., 2002; Rees and Sternberg, 2002; Chentsov, 2004). On different objects the anisotropy effect in microtubules, which is positive in a direction of a spindle (Schmidt, 1939; Inoue, 1953; Swann, 1961) is confirmed. It is suggested, that in the spindle we have to deal with orientated proteins, polarized in the direction of power lines. The electrical field developed by the centrioles, form each molecule of tubulin into the separate charges (opposite by sign and equal by value) during the prophase. During metaphase the orientation of molecules is optic shown in the form of the threads of the achromatic spindle. The tubulin polarization is weaker after the disappearance of the electrical field (telophase) and the spindle breaks down.

In metaphase the chromosomes are built on an equatorial plane and cease to move. They have two chromatids that are identical to the DNA-particle and the quantity of elastic energy enclosed in them. The shoulders of the chromosomes are completely separated. The chromatids are connected only in the centromere region. As the centromere is attracted to two positively charged centrioles, its electrons redistribute and concentrate at opposite parts of the sphere and that leads to the centromere and chromosome splitting.

After a DNA-wave and biofield disappear, the cell becomes a closed system. It is well known, that in all closed systems, entropy never decreases, it either is constant, or it increases. Therefore, at this time in a mother cell, it will spontaneously follow only those processes which transfer it from a more ordered condition to a less ordered one. In prophase the contents of living cells act as if boiling (Haggis et al., 1965). The biomolecules lose the orientation and begin to move casually. Chaotic hits against an envelope change the cells form. The cytoplasm circulation ceases. The nucleolus and the majority of polyribosomes are loosened and disintegrated (Scharff and Robbins, 1966; DiMario, 2004). RNA molecules are transferred into the nucleoplasm and after that into the cytoplasm (Love, 1964; Scharff and Robbins, 1966; McIntosh and Landis, 1974; Strochkova and Kazan'ev, 1975). Left in a cytoplasm, is about 95% of the nuclear proteins (Chapman et al., 1971; Abramova and Neytakh, 1973). Photosynthesis, respiration, growth, intake of nutritious substance and synthetic processes partially or completely stop (Mazia, 1961; Robbins and Scharff, 1966; Epifanova, 1973, 2003; Mitchison, 1973; Strochkova and Kazan'ev, 1975; Ivanov, 1990). ATP quantity reaches a minimum (Mazia, 1961; Chin and Bernstein, 1969). Connection with an environment takes place by a gradient of concentration. As a consequence, the viscosity of protoplasm begins to fall (Heilbrunn and Wilson, 1948; Heilbrunn, 1956, 1958). The ionic balance changes by the gradient of concentration as well: concentration of ions K^{+} and Ca^{2+} decreases, and ions Na^{+}, Mg^{++} becomes higher (Clarckson, 1974; Sorokina, 1978; Kazmin, 1984; Ivanov, 1990). These facts are demonstrated, that a dividing cell is a not working cell (Peter, 1970). It is characterized by an entropy maximum, aspires to equilibrium and a uniform distribution of molecules on all cell volume. It is the interval (2-4 % of all cell cycle), when the cell is "out of a life", as when the life of a mother cell has finished and the life of the daughter cells yet has not begun.

Thus, the 2nd cycle quarter, having begun with DNA-wave and biofield destruction, has ended by the aligning of doubled chromosomes and reached the maximum condensation of chromosomes in an equatorial plane. DNA exists as a two DNA-particles and without a DNA-wave. At this time the cell is a closed system. Happening in it simultaneously is two opposite directed process – a creation and destruction. The first of them proceed as a physical process. It includes the stages of metaphase chromosome formation and the stages of energy and the equal distribution of genetic information. Exactness of mitosis mechanism realization is secured by the closed system conditions. In a mother cell since a prophase, intracellular processes that go in a direction of entropy increase. Changes have an irreversible character. When the energy entrance ceases, the metabolism and cell structure is broken. Connection with an environment is realized on a concentration gradient and is regulated by a cell membrane.

III Cycle Quarter – The Stage of Cell Birth (1 DNA-Particle – 0 DNA-Waves)

In anaphase, the two groups of chromosomes begin a motion under the action of the electrostatic forces of attraction between the positively charged centrioles and negatively charged centromeres (Figure 7d). There it moves slowly, at a velocity of ≈ 0.5-2 mic/min. (Chentsov, 2004), as a gel consistency of nucleoplasm and big "tails" – the chromosome arms, which do not have a charge and are inert. As the electrons are concentrated in a centromere, chromosomes in motion become the V- and J-shaped form. During the approach of chromosomes to centrioles their own centrioles are diverged, taking away with them the chromosomes.

According to the law of charge preservation the charges sum of the closed system is constant. When the centrioles and a set of chromosomes are united, their charges are neutralized (Figure 7d). The motion of the V- shape set of chromosomes occurs in the uniform narrowed ring of centromeres (Bondareva, and Mosolov, 1973). The centromeres at a blow "break" centrioles (a molecule of actin polymer) into the initial number of actin monomers. As consequence in the telophase the SH-group content is increased (Sakai and Dan, 1959; Mazia, 1961; Alov and Aspiz, 1966).

Firstly a periodic change of SH-groups (Rapkine cycle) in cell cycle as described by L. Rapkine (1931). Later on it has been determined, that the substance subject to fluctuations has a protein nature (Gawadi, 1971). A direct relationship between the fluctuations of the protein of thiol groups and formation of the mitotic apparatus is shown (Sakai, H. and Dan, 1959; Mazia, 1958, 1959, 1961; Alov and Aspiz, 1966; Alov et al., 1969). "Thiol paradox" is related in our model with the action of pair structure – centrioles, where they function in the "polymerization - depolymerization" system on the bases of the actin protein.

At this moment the existence of a structure with a charge transfer has ended, becoming DNA-wave radiation. The given structure has a guiding value for proliferate cell development. According to A. Szent-Györgyi (1960), all life on the Earth is set in motion by means of mechanisms, in which the energy of the increased electrons is used. Perhaps, the initiation of photosynthesis and the new cell reproduction process in the basic sense are similar enough – a transmission of increased electron energy on a chain of special molecule-transfers.

In our model the actin molecules serve for catching the radiating DNA-wave energy as a "trap". The electrons, which have absorbed this energy, pass from the basic energy condition into an increased condition. Then energy excitement (i.e., the increased electrons) moves from actin to the DNA-polymerase, transferring it to an active condition. The active enzyme is connected to the DNA. Electrons start to make, at this time, quantum transitions to the basic level, radiating the energy component. The DNA-polymerase transfers the energy of the chemical bonds, catalyzing the DNA biosynthesis reaction by a complementary principle. The non-excited electrons, after replication, remain on the coiling DNA. But after chromosomes arms lose a dipolar design, they are moved down to the centromeres where they carry out their last mission – full chromatid division and delivery of chromosomes to areas of new cell formations.

After the electrical charges neutralize the electric field, which retained the metaphase chromosomes in the condensed condition, they then disappear, and according to its elastic properties, the hard construction of macromolecule begins to uncoil and extend. At uncoiling the position and orientation of the chromosomes, where the centromere regions are directed to a pole and telomere is directed to the spindle center, are not changed (Monachova, 1990, 2004; Chentsov, 2004). Therefore, each chromosome occupies the space inside, organizing the nuclei into their own niche. The uncoiling brings about the destruction of complex isomers and the restoration of the initial dipolar design and configuration of the DNA construction. The given process goes with the liberation of free energy. As a consequence the environment in a cell changes once more.

At the decondensation of chromosomes and at the restitution of DNA dipolar structure, the electrical charges are moved. Their motion, by the laws of physics, generates the electromagnetic field. Therefore, around two groups of chromosomes, regularly uncoiling and quickly increasing in volume, there are the electromagnetic fields. A cytoplasm is attracted to the two field sources. At the beginning the diameter of the fields with a cytoplasm is not big and they, not being covered, keep within a mother cell. However the enlargement of the chromosomes volume leads to an expansion of their fields. At their approach to each other, they, mutually repulsing, move apart the two "balls" of uncoiling chromosomes with a cytoplasm. This sequence of the onset of cytokinesis conforms to the experimental data as well (Mazia, 1961).

The majority of cell molecules are molecular dipoles. Therefore, having reached the zone of an electromagnetic field, they react to it. The field forces them, depending on the electrochemical construction of the molecules, to a different speed. The increase in speed forms a cytoplasm around the uncoiling chromosomes. So, in the telophase, a centripetal motion of many molecules is found (Heilbrunn and Wilson, 1948; Heilbrunn, 1956, 1958; Brachet, 1957, 1960; Neyfakh and Kostomarova, 1971; Stevens and Prescott, 1971; Simmons et al., 1972, 1973; Epifanova, 1973; Epifanova et al., 1977; Yudin and Neyfakh, 1973; Strochkova and Kazan'ev, 1975; Dundr et al., 2000; Cheutin et al., 2004). Due to a centripetal motion of water molecules from an external environment into a cell, the viscosity of the cytoplasm increases again. The RNA molecules appear again in the region of the future nucleus, forming numerous pronuclei. From cytoplasm the component part of proteins come back. With a centripetal motion, the phospholipids restore a nucleus envelope. However, the speed of pectic substances, apparently, is not great. They accumulate on the periphery and are exported for use as the boundary of the future cells (Swanson and Webster, 1977; Dorochova and Shanina, 2004, 2005). Pectic substances, fusing together, give rise to a mid-body – an amorphous layer which surrounds the daughter cells, promoting, thus, in their isolation. Apparently, the electromagnetic field destructs products of metabolism of mother cell as well.

The cell is a closed system. Its main processes, such as respiration, photosynthesis, RNA, protein synthesis, etc. either are absent, or are supported at a low level (Mazia, 1961; Epifanova, 1973, 2003; Smolenskay, 1973; Zosimovskay, 1973). At telophase the summary weight of daughter cells does not exceed the maternal cells weight. The total RNA and protein quantities in each of them are at half.

In a closed system, if the body does not act on the external forces, the sum of the kinetic and potential energy ($E_{kin.}$ + U) must be constant. Chromosomes decondensation takes place during the telophase. Accordingly, their potential energy decreases and the kinetic energy of the telophase cells increases. The uncoiling chromosomes include two consecutive processes. At the onset, the threads spiral, packed by external coiling, and only then by the internal coiling are they untwisted. These processes go along with the secretion of free energy. The potential energy, which is concentrated into metaphase chromosomes, is once more (2nd) the internal energy source of a proliferate cell.

Under the first thermodynamic law of a closed system, this energy is preserved, increasing the energetic state of the daughter cells at the end of the telophase. According to experimental data the free energy from the 2nd internal source is utilized by the five main channels (Figure 8).

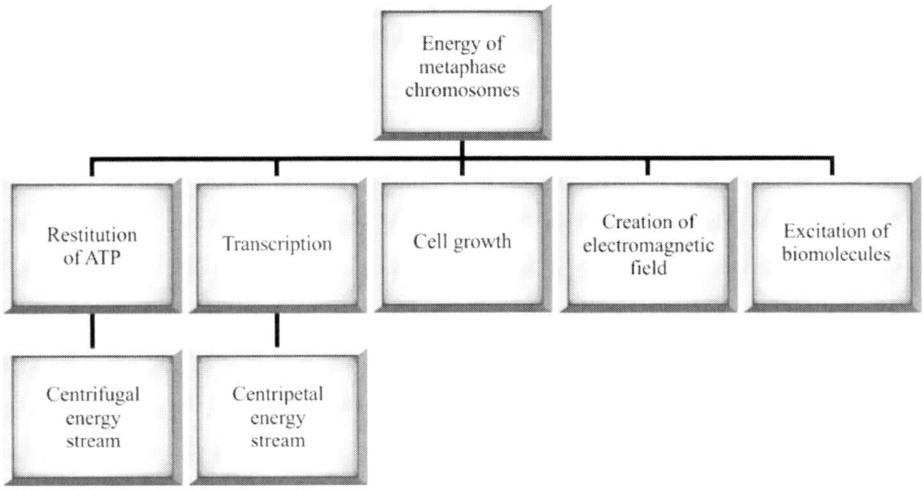

Figure 8. The scheme of utilization energy from metaphase chromosomes.

One energy part will absorb the RNA-polymerase, transferring it in an active condition. It is established, that during the telophase the intensive process of RNA synthesis occurs (Swanson and Webster, 1980; Alberts et al., 2002, etc). RNA biosynthesis proceeds exclusively on the euchromatin regions of DNA. The second part – will go on reconstruction of the ATP, carrying out in the cell a role of the preserver and conductor of energy. It is established, that during cytokinesis the ATP quantity increases (Mazia, 1961). The third part will increase the kinetic energy of the molecules, causing an increase of intracellular pressure. The fourth part will use an electromagnetic field creation to which the function of the DNA-wave and order of nuclear and cytoplasmic function will be restored. Lastly, the fifth part of the energy will go on excitation of molecules and will be revealed in the form of fluorescent radiation (Gurvich, 1932, 1944).

The cells at the second and third cycle quarter have many common properties. These are the closed systems, vital functions are practically absent, and mitosis passes the determinate stages of division. But the maternal cell (the second cycle quarter) constitutes a colloidal solution, in which the molecules are in chaotic motion and cell content combines. The daughter cells (the third cycle quarter) are orderly and territorially isolated formations. Their energy is restored, and the disposition of the molecules is regulated. Their ontogeny only needed to be "switched on".

Thus, the third cycle quarter, having begun with a chromosome separation, the formation of daughter cells in a volume equal to the maternal cells is finished (Figure 7f). DNA exists in the form of one DNA-particle and without a DNA-wave. The cell is a closed system. Respiration, photosynthesis and synthetic processes almost are absent. But the energy which has entered from the second internal source, and the electromagnetic field which has sorted mother cell contents into equal parts, have prepared an energy and material basis for daughter cells to transition into an ontogeny.

IV Cycle Quarter – The Stage of Ontogenesis (1 DNA-Particle – 1 DNA-Wave + 2 DNA-Particles – 1 DNA-Wave)

The term "ontogeny" (growth) for cells was introduced by B.P. Tokin in 1934, during a study of the life cycle of Protozoa cells. The period of cell ontogeny, which enlarges upon the end of the interphase, begins at the moment of the formation of a nucleus, when uncoiling chromosomes hit against the envelope on the centromere and two telomere regions and it comes to an end during early prophase. This blow causes a reaction which once again changes a cell's environment. 1. The macromolecule completely uncoils by the destruction of complex isomers and acquires the most typical form - a double spiral. 2. On the DNA a standing wave forms and it starts to exist as 1 DNA-particle and 1 DNA-wave. 3. The DNA-wave becomes a biofield source, which descends as 1/r. 4. The cell transforms to an open and quantum-mechanical system. The development of cell begins to carry out in the form of the two-plane (corpuscular-wave) biophysical process, which was named as the stage of ontogenesis.

The portion of energy from the second internal source, as was mentioned above will go on an increase of osmotic pressure into a newly formed cell. The daughter cells, which have begun biosynthesis, deposit a layer on the primary wall (Swanson and Webster, 1977; Esau, 1977; Chentsov, 1978, 2004). The tension of the membrane under hydrostatic pressure begins to grow, a thin and elastic primary wall is stretched, and a cell increases in volume. The growth by stretching occurs during G_1 period up to the formation of the secondary cell wall (Anderson et al., 1969; Chentsov, 1978, 2004, Geitmann and Ortega, 2009). Every daughter cell receives from the metaphase chromosomes, the same energy quantity. That is why the growth of the cytoplasm is limited by specific boundaries known as the nucleocytoplasmic ratio and the volume of the cell during the interphase is doubled.

A study of RNA dynamics during the mitotic phases has established, that their quantity decreases in the prophase and again increases in the telophase of mitosis (Prescott and Goldstein, 1968; Stevens and Prescott, 1971; Prescott, 1970; Mitchison, 1973; Smolenskay, 1973). The molecules of the RNA-polymerase, which has received energy from the second internal source, pass into the active form. All enzymes are molecular dipoles and have own electrical field. Therefore they are attracted to one of DNA threads (the complimentary interaction of two polar molecules), forming a stable connection (Georgyev, 1968; Rees and

Sternberg, 2002). In the places of their attachment the asymmetry of charges increases and causes a reduction of hydrogen connections and a divergence of the threads of DNA.

A synthesis of 70-85 % of all rRNA is carried out in the nucleolar organizing region (Georgyev, 1968; Swanson and Webster, 1977; Chelidze, 1985; Lehninger, 1985; Alberts et al., 2002; Chentsov, 2004, etc.). The nucleolar organizer is a segment of chromosomal DNA located, as a rule, in the region of the secondary constriction of the chromosome. The working zone in it (the transcript unit) represents a decondensed polyisogenic fragment of the DNA, limited to heterochromatin blocks. One matrix of one transcript unit can simultaneously synthesize two hundred rRNA molecules (Miller and Beatty, 1969). Biosynthesis of rRNA begins in a place where the heterochromatin block is changed on a euchromatin area. In our judgment, such localization and construction of a copied region provides success in all stages of transcription – initiation, rRNA biosynthesis and termination, as chemical processes are catalyzed by an enzyme of the RNA-polymerase.

In our model, the processes transcription and replication are rather similar. Having connected to a strand of DNA and moving with its wave, the enzyme simultaneously untwists into a double spiral DNA-particle and builds a new RNA molecule on a counter wave of the DNA-wave. However, the electrical field of the RNA-polymerase is weaker than that of a DNA-polymerase. It is created only because of the electrical charges asymmetry. Therefore this enzyme can untwist and, as consequence, transcribe only euchromatin or decondensed regions, which the number of these on the DNA-particle is limited. The correlation between the degree of chromatin condensation and its transcript activity is confirmed many times. In the end of biosynthesis an inactive enzyme and a new rRNA molecule collide with the second heterochromatin block and separate from the DNA.

The biosynthesis of new generations of rRNA molecules begins at the moment of DNA-wave rise. In the beginning they are formed in the appearance of big primary transcripts – the rRNA predecessor (pre-rRNA) with a sedimentation coefficient nearing 45S (Georgyev, 1968; Stevens and Prescott, 1971; Kiknadze, 1972; Swanson and Webster, 1977; Lehninger, 1985; Alberts et al., 2002; Andersen et al., 2005; etc.). The place of their localization becomes pre-nucleolus, consisting in an early telophase from the RNA molecules of the old generation. At this moment the pre-nucleolus are reconstructed in the in depend of structure of a nucleus – a nucleolus. In the nucleolus all pre-rRNA is accumulated. After rRNA biosynthesis termination the nucleolus loses connection with the DNA. A stereoscope study of the nucleus has shown, that in a telophase their body is narrow, closed with chromosomes, in a late telophase they move away and the distance between them reaches a maximum in an interphase (Bondareva and Mosolov, 1975).

Processing is a multi-staged function of the RNA and its ordered disintegration that is constantly going in the nucleus, nucleolus, ribosome's and cytoplasm of the cells. It has been proven that, soon after formation, all heavy pre-rRNA disappears (Georgyev, 1968; Lehninger, 1985; Dundr et al., 1996; Cheutin et al., 2004; Chentsov, 2004; DiMaria, 2004; Andersen et al., 2005). In its place there are the same fragments of rRNA in these sizes, 28S, 18S and 5,8S. At splitting, pre-rRNA, fragments and is decreased to half of a molecule. 50 % of typical rRNA (28S and 18S) is deconstructed, not leaving a nucleus. During interphase the pool of pre-rRNA is gradually exhausted and in the metaphase a greater portion of them is deconstructed. Nevertheless during mitosis the remains of the processed pre-rRNA are still preserved. The greater part of early mRNA and tRNA is subjected to a processing as well. A disintegrating RNA is no different from RNA, which has not been exposed to disintegration.

Most likely, processing is the physical phenomenon of exponential disintegration. The phenomenon of exponential disintegration is often found in nature (heavy nuclei α-disintegration, disintegration U^{238} etc.). Its essence: at the spontaneous division of particles the probability of their disintegration never changes, therefore by the end of the first half of the disintegration period there remains 1/2 particles, the second – ¼, etc. The process of disintegration always goes with the liberation of free energy. Therefore we are considering the phenomenon of processing as a centrifugal energy stream prolonged in time (Figure 9). Cells use the energy from this stream through definitive channels. It is necessary for it to support a working condition during interphase. In the foundation of the given stream is the energy liberated by the second internal source. The energy entering into a nucleus from the metaphase chromosomes during their uncoiling is reserved in the chemical connections of various RNA molecule-precursors. Then, during their maturation or the exponential disintegration on the intermediate (41S-; 32S-; 20S-) and final fragments (28S-; 18S-; 5,8S-) of the rRNA, this energy is liberated, and in this way the energy balance of the interphase cell is constantly supplied. This energy stream is internal by origin, and the vector of energy motion is directed from the metaphase chromosomes to the cytoplasm. That is why it is designated as centrifugal. The rRNA has the leading role in the centrifugal energy stream, as most often meeting the RNA form. It is well known, that all outgoing transcripts of rRNA molecules-precursors are identical on a chemical compound basis and on a polymer length basis. The requirement is that the processing must go according to the determined scheme, and the centrifugal power stream is a strictly regulated process in time, and in this way the power balance of a cell is constantly supported.

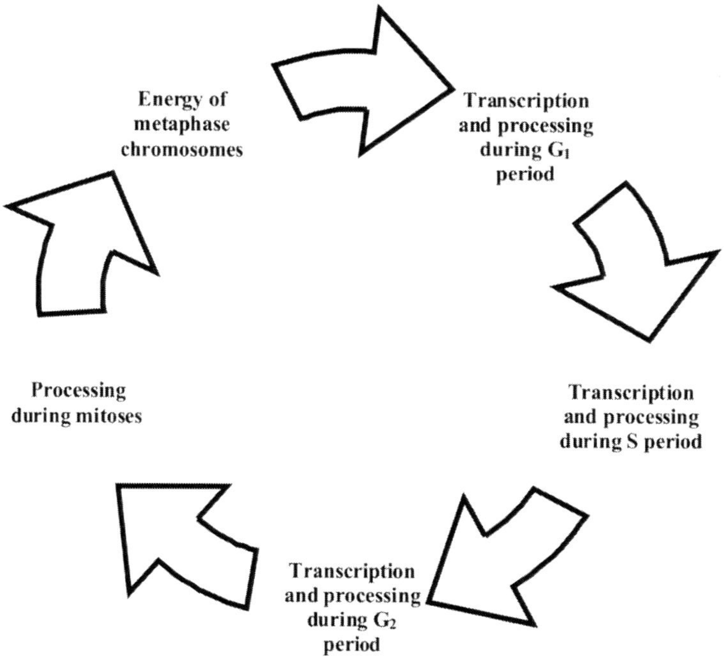

Figure 9. Centrifugal energy stream on the basis of the exponential disintegration of RNA molecules or the phenomenon of processing.

Having analyzed the experimental data, we have come to the conclusion that the basic function of a centrifugal stream is to provide energy by protein biosynthesis . The place of protein biosynthesis is in the ribosomes. They are found in almost all the cells, in the cytoplasm and nucleoplasm, in free condition or connected with a membrane. We propose that in the translation processes all RNA kinds are involved. The mRNA appears as a matrix of genetic information, the tRNA as a transport system and reading system, and the rRNA as an energy reservoir. It is no accident that the rRNA makes up half the mass of the ribosomes. Their work is organized so, that up to interphase end, the protein quantity in the cell is doubled (Smolenskay, 1973).

Exact spatial localization of intracellular processes and structures is an important feature of eukaryotic cells. The place and time of the biochemical reactions is strictly compartmentalized. The essential influence on the micro heterogeneity and on the organization of cell internal space is a biofield and the decreasing intensity of the gradient which is directed from a nucleus to the cytoplasm. It prevents chaotic (thermal) molecule motion in a cell. The forces of a biofield on dipolar molecules and ions, give different kinetic energies, causing vector movement or a stop in a nucleus or cytoplasm. The result of such self-regulation is the presence of the ordered streams of energy and substance in a cell.

Studying of Escherichia coli behavior in substrate hunger conditions has shown, that in the absence of nutrition, the population of microorganisms has survived for as long as 2 years and has preserved the ability of reproduction (Gusev, 2001). Hence, the energy quantity entering into a cell from a DNA-wave is rather great. But this energy is not the same that the cell population would have normally developed, reproduced and for the work in different environment conditions. For this it would require the entrance of energy from the outside.

The eukaryotic cell by its structural and functional attitude is organized in a more complicated manner, than the prokaryotic cell. Therefore for the maintenance of all intracellular systems and processes in a viable condition it would be necessary to have much more energy. i.e., the life of eukaryotic cells is directly dependent on the presence of nutrients in the environment. Cells that are deprived of nutrients from the outside are stopped in their development during the G_1 period. The speed of the entering exosubstrate into the cell is increased sharply from the beginning of the interphase, and soon reaches a maximum (Ivanov, 1990). Simultaneously, the restoration of the ionic composition is occurring. As mentioned above, in dividing cells (i.e., when there is no DNA-wave and biofield) the concentration of Na^+ increases, and K^+ decreases. After the chromosomes decondensation, the DNA-wave and biofield restoration, K^+ ions begin to enter into the cell and Na^+ transport is directed outward from the cytoplasm (Clarckson, 1974; Kazmin, 1984; Rubin, 1999). Apparently, the process of ion active transport occurs with the participation of biofield forces.

Energy processes in plants are similar enough. All of them are aerobe (requiring oxygen). The sources of outer energy are of two types – photosynthesis and respiration. Both of them are acting only during the interphase when the cell is an open system. Their work is connected with well-regulated biophysical and biochemical processes of chemical energy and light energy transformations, with participation defined organelles (chloroplasts, mitochondria's, etc.). The end product of these two types of reactions is a molecule ATP. Thus, the second energy canal, which takes part in the cell energetics during the interphase, is a centripetal stream, which operates in the system "ATP-ADP" (Figure 10). Three energy sources are involved in its work – photosynthesis, respiration and metaphase chromosomes. In the telophase the 2nd energy source creates a centripetal stream, and two external sources support it during the interphase. The ATP molecules arising in the course of photosynthesis and respiration have increased their intracellular supply that has prevented a sharp drop in its level in the interphase cell.

At the beginning of interphase the ultrastructure of the cytoplasm of the daughter cells is reformed in accordance with the state of their genome (i.e., the state of their DNA-particle and DNA-wave). An example of *Coleps hirtus* infusorian showed that during the two hours after the division takes place, a restoration of the intracellular structures of the daughter cells that were destroyed during the division (Kalinnikova and Severtsov, 1983). Their energy is added from the two energy streams – centripetal (in system "ATP-ADP") and centrifugal (on the basis of processing). The centrifugal stream is formed at the expense of energy liberated during the exponential disintegration of RNA molecules. Therefore, the volume and decrease of the given stream are defined by the number of the synthesized molecules pre-rRNA and the limit of their half-life periods. The minimum level of a centrifugal stream is reached at the ontogeny end (and the next prophase beginning). According to experimental data, the intensity of the processes of photosynthesis and respiration and, accordingly, the ATP level (a centripetal stream) falls to a minimum at the same time (Mazia, 1961; Bell, 1980; Selichanova and Zalenskiy, 1982). As known, without energy, the cell is doomed to death. On different species of Protozoa it has been determined that before cell division, the intensity of metabolic processes decreases, the intensity of respiration and protein biosynthesis changes the reaction to poisons, vital dyes, etc. (Sarkisov and Vtyurin, 1967; Tokin, 1987).

Death overcomes the next DNA-wave of radiation. Initiated by its energy the development of new cells approaches the point of irreversibility to the end of the G_2 period, when the mechanism of cell doubling is impossible to stop. Therefore the cell cycles in the two adjacent generations of cells (mother and daughter) are covered again, from the interphase S-period up to the middle prophase. Then the mother cell dies, and instead of it there are two daughters, which after cytokinesis begin an in depend of life.

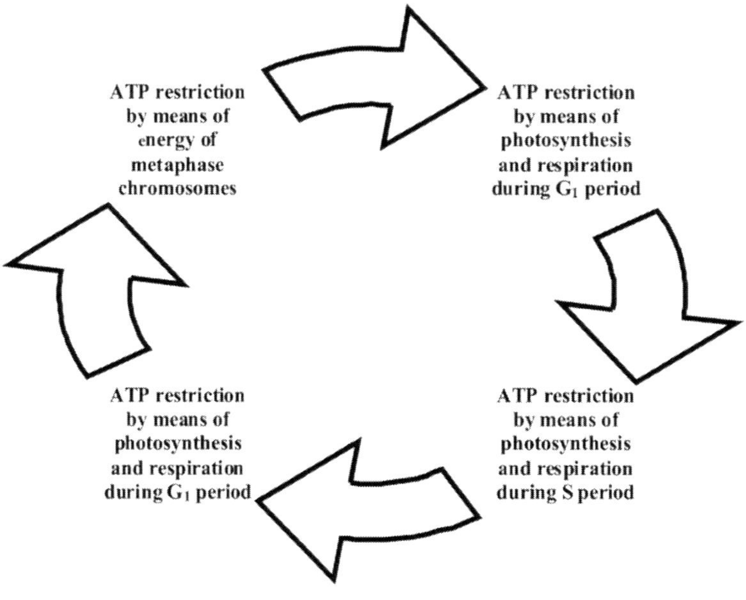

Figure 10. Centripetal energy streams on the basis of the energy of ATP molecule.

Conclusion

In corpuscular-wave model the full life cycle of proliferate plant cells begins from S period and continues until the next prophase. In the initial development of the cells of a new generation a DNA-wave transitions to the lower power level, and enters into a cell of free energy quantum in the end of G_1 period participates. The initial stages of their formation (DNA replication, chromosomes coiling, centrioles formation) takes place in an interphase mother cell, in which ultrastructure and viability processes after DNA-article doubling are not changed, and the cell is an open system. The stages connected with mitotic development and cell birth (prophase, metaphase, anaphase and telophase mitosis) proceed into a mother cell as well, which becomes a closed system. At this time its biological harmonious oscillator, internal organization and synthetic processes are destroyed, and the cell is characterized by a maximum of entropy. The in depend of life cells (ontogeny) begin from cytokinesis, restoration of nucleus envelope, "switching on" of the biological harmonious oscillator, reconstruction of nucleus and cytoplasm orderliness, appearance of centrifugal and centripetal energy streams, the cell transition from a closed to an open system occurs somewhere near the telophase end. The ontogeny period goes from beginning to end interphase and finishes by cell death. During cell cycle a DNA undergoes the following changes: {2 DNA-particles–1 DNA-wave} → {2 DNA-particles–0 DNA-wave} → {1 DNA-particle–0 DNA-wave} → {1 DNA-particle–1 DNA-wave + 2 DNA-particles–1 DNA-wave}. For more than 90% of a cell's cycle the cell develops as an open system (interphase). The energetics is provided by two internal and two external in depend of energy sources (DNA-wave, metaphase chromosomes, respiration and photosynthesis). The cell cycle mechanism is constructed so, that on its end, daughter cells have the same DNA-particles of a mother cell and a DNA-wave in which the energy level is less, than at a mother cell on one level.

One of the biggest jumps of evolution is the emergence of eukaryotic cells (Margulis, 1970; Pickett-Heaps, 1974; Swanson and Webster, 1977; Mednikov, 1982 Stack and Anderson, 2001). The evolutionary alterations also affected the construction of the biological harmonic oscillator. It shifted territorially from the cell into the nucleus. This essentially increased its safety. The energy capacity of a harmonic oscillator is increased by means of the growth in chromosome numbers connected with the nucleus envelope by three points – in the centromere and two telomere regions. This gives the stimulus for the further development of the genome by the corpuscular and wave it constitutes. The DNA-wave begins to take possession of the higher levels of energy ($E = 1, 2, 3 \ldots n$), as a result a living matter could pass to a multicellular level of organization.

One more evolutionary novelty was the emergence of an additional (2nd) source of internal energy - metaphase chromosomes. The energy distribution in cells is not random. A given source together with a biological harmonic oscillator is regulating during the cell cycle, the energy expenditure between the two major intracellular processes - replication and transcription. Condensation and decondensation of chromosomes is a reversible process. Energy is accumulated in the course of chromosome coiling during the G_2 period, prophase and metaphase and is liberated when they are uncoiling during telophase. The DNA-wave, DNA-particle and two centrioles take part in the formation of metaphase chromosomes and after the splitting and divergence of the chromosome set, the daughter cells receive with respect to the mother cell, an identical DNA-particle, and a DNA-wave down one level and the same quantity of free energy.

References

Abramova, N.B. and Neytakh, A.A. (1973). Migration of newly synthesized RNA during mitosis. 2I. Nuclear RNA in the cytoplasm of metaphase cells. *Ibid.*, 77(1-2), 136-142.

Actin: A dynamic framework for multiple plant cell function. (2000). Eds. Staiger C. et al. Dordrecht., Kluwer.

Alberts, B., Johnson, A., Lewis, J., Raff, M., Roberts, K. and Walter, P. (2002). *Molecular biology of the cell*, 4th edn. Chaffey, N (Ed.), Garland Science, New York.

Alieva, I.B., Borisi, G.G. and Vorobyov I.A. (2008). Spatial organization of microtubules associated with centrosome in the cytoplasm of fibroblasts ZTZ. *Tsitology*, 50(11), 936-947.

Allen, J.P. (2008). *Biophysical chemistry*. Wiley-Blackwell Publ., Oxford.

Alov, I.A. and Aspiz, M.E. (1966). The value of sulfhydryl groups of protein in the process of mitotic cell division. *DAN SSSR*, 166(4), 965-967.

Alov, I.A. (1972). *Cytophysiology and pathology of mitosis*. Moscow, Medicina.

Alov, I.A. (1975). Preprophase. *V. 2. Cytology. Mitotic cell division.* pp. 67-89. Moscow, VINITI AN SSSR.

Alov, I.A., Braude, A.N. and Aspiz, M.E. (1969). Basics of cell functional morphology. Moscow, Medicina.

Andersen, J.S., Lam, Y.W., Leung, A.K., Ong, S.E., Lyon, C.E., Lamond, A.I. and Mann M. (2005). Nucleolar proteome dynamics. *Nature,* 433, 77-83.

Anderson, E.C., Bell, G., Petersen, D. and Torbey, R. (1969). Cell growth and division. 4. Determination of volume growth rate and division probability. *Biophys.*, 9(2), 246-263.

Aspiz, M.E. (1975). Mitotic apparatus. *V. 2. Cytology. Mitotic cell division.* pp. 120-150. Moscow, VINITI AN SSSR.

Bell, L.N. (1980). *Energeticsof the photosynthesizingplant cells*. Moscow, Nauka.

Bondareva, F.F. and Mosolov A.N. (1975). Stereoscopic study of chromosome disposition in cell cycle. *Cytology and Genetic*, IX(2), 99-100.

Brachet, J. (1957). *Biochemical cytology*. Academic Press, New York.

Brachet, J. (1960). *The biochemistry of development*. Pergamon Press, London, New York, Paris, Los Angeles.

Cantor, C.R. and Schimmel, P.P., 1980. *Biophysical chemistry*. Freeman, W.H. (Ed.), San Francisco.

Cell cycle control. (1995). Eds. C. Hutchison, D.M. Glover. New York, Oxford Univ. Press.

Chapman, J.D., Webb, R.G. and Borsa, J. (1971). ATP pool levels in synchronously growing Chinese hamster cells. *Cell Biol.*, 49(1), 229-337.

Chelidze, P.V. (1985). *Ultrastructure and function of the nucleolus of interphase cell*. Tbilisi, Metsniepeba.

Chentsov, Yu.S. (1978). *General cytology*. Moscow, Moskovvski universitet.

Chentsov, Yu.S. (2004). *Introduction to cell biology*. Akademkniga, Moscow.

Cheutin, T., Misteli, T. and Dundr, M. (2004). *Dynamics of nuclear components*. In: The nucleolus. New York, Kluwer Academic, Plenium Publ., 29-40.

Chin, B. and Bernstein, I.A. (1969) Adenosine triphosphate and synchronous mitosis in *Physarum polycephanum*. *Cell Biol.*, 96(2), 330-337.

Clarckson, D. (1974). *Ion transport and cell structure in plants*. McGraw-Hill, London, UK.

Cremer, T., Kreth, G. and Koester, H. (2000). Chromosome territories, interchromatin domain compartment and nuclear matrix: an integrated view of the functional nuclear architecture. *Crit. Rev. Eukaryot. Gene Expr.*, 10, 179-212.

DiMario P.J. (2004). Cell and molecular biology of nuclear assembly and disassembly. *Int. Rev. Cytol.,* 239, 99-178.

Dorochova, N.V. and Shanina, N.V. (2004). The cycle of microtubulecyto skeleton in plant cell division. *Tsitology*,46(11), 960-966.

Dorochova, N.V. and Shanina, N.V. (2005). Organization of cytokinesis in plant cell division. *Tsitology*, 47(7), 563-577.

Drake, J.W., Allen, E.F., Fosberg, S.A., Preparata, R.M. and Greening, E.O. (1969). Genetic control of mutation tastes in bacteriophase T4. *Nature*, 221, 1128-1132.

Dundr, M. and Olson, M.O. (1998). Partially processed pre-rRNA is preserved in association with processing components in nucleolus-derived foci during mitosis. *Mol. Biol.Cell.* 9, 2407-2422.

Dundr, M., Misteli, T. and Olson, M.O. (2000). The dynamic of postmitotic reassembly of the nucleolus. *J. Cell Biol.* 150,

Edenberg, H.J. and Huberman, J.A. (1975). Eukaryotic chromosome replication. *An. Rev. Genet.*, 9, 245-284.

Epifanova, O.I. (1973). Studies on the control mechanisms of the mitotic cycle using inhibitors of transcription and translation. *The cell cycle. Problems of regulation.* Moscow, Nauka, 72-103.

Epifanova, O.I. (2003). *Lectures about a cell cycle*. Moskow, KMK.

Epifanova, O.I., Terskih, V.V. and Zaharov, A.V. (1977). *Radioautography*. Moscow, Visshaya shkola.

Esau, K. (1977). *Anatomy of seed plants*. J. Wiley and Sons, Inc., New York, Santa Barbara, London, Sydney, Toronto.

Feynman, R.P., Leighton, R.B. and Sands, M. (1963). *The Feynman Lectures of Physics*. vol. 1-3. Wesley, A. (Ed.), Readigg, Massachusetts, Palo, Alto, London.

Friedman, D.L. (1970). DNA polymerase from HeLa cell nuclei: levels of activity during synchronized cell cycles. *Biochem., Biophys. Res. Com.*, 39(1), 100-109.

Gawadi, N. (1971). Actin in the mitotic spindle. *Nature*, 234(5329), 410.

Geitmann, A. and Ortega, J.K.E. (2009). Mechanisms and modeling of plant cell growth. *Trends of Plant Science,* 14(9), 467-478.

Georgyev, G.P. (1968). The nucleus. *Enzyme cytology*. Ed. D.B. Roodyn, Academic Press, London, New York.

Gurvich, A.G. (1932). *Mitogenetic radiation*. Medgiz, Moscow.

Gurvich, A.G. (1944). *Theory of biological field*. Sovetskay nauka, Moscow.

Gusev, V.A. (2001). About the energy source for preserve the viability and amplification of heterotrophic organisms in the conditions of substrate hunger.I. Formulation of hypothesis. *Russian J. Biophysics*, 46(3), 862-868.

Haggis, G.N., Michel, D., Muir, A.B., Roberts K.B. and Wolter P.M.R. (1965). *Introduction to molecular biology*. Wiley, J. (Ed.), New York.

Hand, R. (1978). Eukaryotic DNA: organization of the genome for replication. *Cell*, 15, 317-325.

Harland, J., Jackson, J.F. and Yeoman, M.M. (1973). Changes in some enzymes involved in DNA biosynthesis following induction of division in cultured plant cells. *Cell Sci.*, 13(1), 121-138.

Heilbrunn, L.V and Wilson, W.L. (1948). Protoplasmic viscosity during mitosis in the egg of *Chaetopterus. Biol. Bull.*, 95(1), 57-68.

Heilbrunn, L.V. (1956). *The dynamics of living protoplasm.* New York, Academic Press.

Heilbrunn, L.V. (1958). The viscosity of protoplasm. *Protoplasmatology. 2*, Heilbrunn, L.V. and Weber F. eds., Springer-Verlap, Vienna.

Inoue, S. (1953). Polarization optical studies of the mitotic spindle fibers in living cells. *Chromosoma*, 3(5), 487-500.

Ishizaki, T., Morishima, Y., Okamoto, M., Furuyashiki, T. and Narumiya S. (2001). Coordination of microtubules and the actin cytoskeleton by the Rho effector mDial. *Nat. Cell Biol.*, 3, 8-14.

Ivanov, V.N. (1990). Alteration of exo- and endotrophy in the mitotic cell cycle. *Uspekhi sovremennoi biolog2,*110(5), 284-289.

Kalinnikova, V.D. and Severtsov, A.S. (1983). The life cycles of parasitic protozoa as a form of ontogenesis. *J. of General Biology*, XIV(2), 108-121.

Karp, G. (1996). *Cell and molecular biology: concepts and experiments.* J. Wiley and Sons, Inc. New York, NY.

Kazmin, S.D. (1984). *Biochemistry of the mitotic cycle of tumor cells.* Kiev, Naukova dumka.

Kiknadze, I.I. (1972). *Functional organization of chromosomes.* Leningrad, Nauka.

Kiseleva, E. and Allen T.D. (2004). Nuclear pores and nuclear matrix: structure and interaction. Genetics in XXI century: current state and prospects of development. *Mat. 21 Congr. VOGiS (Moskow, 6-12 Aug. 2004). V. 2*, p. 215, Moscow.

Kornberg, A. (1980). *DNA replication.* W.H. Freeman (Ed.), San Francisco.

Lark K.G. and Lark C.A. (1979). *RecA*-dependent DNA replication in the absence of protein synthesis: characteristics of a dominant lethal replication mutation, *dnaT* and requirement for *recA* function. *Cell Spring Harbor Symp. Quant. Biol.*, 43, 537-550.

Lehninger, A.L. (1985). *The principles of biochemistry.* V. 1-3. Moscow, Mir.

Love, R. (1964). Studies of the cytochemistry of nucleoproteins. I. Effect of colchicine on the ribonucleoproteins of the cell. *Exp. Cell Res.*, 33(1-2), 216-231.

Manton, I., Kowallic, K. and Stosch, von H. (1969). Observations on the fine structure and development on the spindle at mitosis and meiosis in marine centric diatom *Lithodesmium Undulatum, Microscopy*, 89(3), 295-320.

Margulis, L. (1970). *Origin of eucariotic cells.* Yale University Press, New Haven.

Mazia, A. (1966). As the cell divide. *The living cell.* pp. 67-92. Moscow, Mir.

Mazia, D. (1957). Biophysical and biochemical investigations of cell division. *Questions of biophisics.* pp. 136-204. Moscow, Inostrannaya literature.

Mazia, D. (1958). SH-compounds in mitosis. I. The action of mercaptoethanol on the eggs of the sand dollar Dendraster excentricus. *Exp. Cell Res.*, 14, 486-494.

Mazia, D. (1959). The role of thiol groups in the structure and function of the mitotic apparatus. *Sulfur in Proteins.* pp. 367-389. Acad. Press, New York.

Mazia, D. (1961). *Mitosis and the physiology of cell division.* Brachet, J. and Mirsky, A. (Eds.), Academic Press, New York, London.

McIntosh, J. and Landis, S. (1974). Centriole ultrastructure and its possible role in microtubule formation in a quatic fungus. *Protoplasma*, 80(1-3), 91-108.

Mednikov, B.M. (1982). Regularities ofgenome evolution. *Molecular mechanisms of genetic processes*. pp. 76-85. Moscow, Nauka.

Miller, O.L. and Beatty, B.R. (1969). Visualization of nucleolar genes. *Science*, 164, 955-957.

Mitchison, J.M. (1973). *The biology of the cell cycle*. Cambridge, Univ. press.

Monachova, M.A. (1990). Cytogenetic aspects of space organization of interphase nucleus. *Uspekhi sovremennoi biolog2*, 10(5), 163-179.

Monachova, M.A. (2004). System organization of genotype in the pheno type of cell nucleus. Genetics in XXI century: current state and prospects of development. *Mat. 21 Congr. VOGiS (Moskow, 6-12 Aug. 2004). V. 2*, p. 293. Moscow.

Moritz, M. and Agard, D. (2001). G-tubulin complexes and microtubule nucleation. *Curr. Opin. Struct. Biol.*, 11, 174-181.

Mueller, G.C. and Kajiwara, K. (1966). Early and late replicating deoxyribonucleic acid complexes in HeLa nuclei. *Biochem. and Biophys. Acta.*, 114(1), 108-115.

Murphree, S., Moore, C. and Peterson, D. (1974). Temporal variation of adenine ribonucleotids during the cell cycle of Chinese hamster fibroblasts in culture. *Exp. Cell Res.*, 83(1), 186-190.

Neyfakh, A.A. and Kostomarova, A.A. (1971). Migration of new-synthesized RNA duringmitosis.I.Embryoniccellsof loach. *Tsitology*, 13(2), 231-235.

Okazaki, R., Okazaki, T., Sakabe, K., Sugimoto, K., Kainuma, R., Sagino, A. and Iwarsuki N. (1968). In vitro mechanism of DNA chain growth. *Gold Spring Harbor Symp. Quantivi Bid.*, 33, 129-142.

Paponov, V.D. (1987). The dynamics of eukaryote genetic apparatus at differential transcription of the genome. *Uspekhi sovremennoi biolog2*, 104(5), 167-183.

Pawelets, N., Mazia D. and Finze M.E. (1984). The centrosome cycle in the mitotic cycle of Sea urchin eggs. *Exp. Cell. Res.*, 152, 47-65.

Paweletz, N. (1968). Elektronenmikroskopic der unbreinflubten und beeinflubten mitotischen Zellteilung, *Aktuelle Probl. Geb. Cancerol. 2.* pp. 191-199. Berlin-Heiderlbery-N.Y.

Pederson, T. and Aebi, U. (2005). Nuclear actin extends, with no contraction in sight, *Mol. Biol. Cell.*, 16, 5055-5060.

Peter, K. (1970). Die indikekle Teilung der Zelle in ihren Beziehungen zu Tatigkeit, Differenzierung und wachstum. *Ztschr. Zellforsch.*, 30(5), 721-750.

Pickett-Heaps, J.D. (1974). The evolution of mitosis and the eukaryotic condition. *Biosystems*, 6, 37-48.

Prescott, D.M. and Goldstein, L. (1968). Proteins in nucleocytoplasmic interactions. III. Redistributions of nuclear proteins during and following mitosis in Amoeba proteus. *Cell Biol.*, 39(2), 404-414.

Prescott, D.M. (1970). The structure and replication of eukaryotic chromosomes. *Adv. Cell Biol.* 1, 57-117.

Rapkine, L. (1931). Sur les processes climiques an cours de la division cellulaire. *Ann. Physiol. Physiochim. Biol.*, 7, 382-418.

Rees, A.R. and Sternberg M.J.E. (2002). *Introduction to molecular biology: From cells to atoms*, Moscow, Mir.

Riegel, J.A. (1967). *Energy, life and organism*. Mir, Moscow.

Robbins, F. and Scharff, M.D. (1966). Some macromolecular characteristics of synchronized HeLa cells. *Cell Synchrony*. pp. 353-374. N.Y.-London, Acad. Press.

Rubin, A.B. (1999). *Biophysics*. V. 1-2. Nauka, Moscow.

Sakai, H. and Dan, K. (1959). Studies on sulfhydryl groups during cell division of sea urchin egg. *Exp. Cell Res.*, 16(1), 24-42.

Sarkisov, D.S. and Vtyurin, B.V. (1967). *Electron microscopy of destructive and regenerative intracellular processes*. Moscow, Nauka.

Scharff, M.D. and Robbins E. (1966). Polyribosome desegregation during metaphase. *Science*, 151(3713), 992-995.

Schmidt, W. (1939). Doppelberrechung der Kernspindel und Zugfasertheorie der Chromosomen-bewegung. *Chromosoma*, 1, 253-264.

Selichanova, O.A. and Zalenskiy O.V. (1982). The junction of the processes of photosynthesis and respiration. *Plant physiology*. pp. 130-145. Moscow, Mir.

Shakhbasov, V.G., Shkorbatov, Yu. P., Strashniuk, V.Yu. (1986). Regulation of activity of nuclear genome and the bioelectrical properties of chromatin and cell nucleus. *DAN SSSR*, 290(5), 1255-1259.

Simmons, T., Heywood, P. and Hodge, L.D. (1972). Mitosis in mammalian cells: studies on the resumption of RNA synthesis. *Ibid.*, 55(2), 240-248.

Simmons, T., Heywood, P. and Hodge, L.D. (1973). Nuclear envelope-associated resumption of RNA synthesis in late mitosis of Hela cells. *Ibid.*, 59(1), 150-164.

Smolenskay, I.N. (1973). Role of RNA in the control of mitotic cycle events. *The cell cycle. Problems of regulation*. pp. 38-71. Moscow, Nauka.

Sorokina, Z.A. (1978). Condition of potassium, sodium and water in cell cytoplasm. Kiev, Naukova dumka.

Sounson, K., Merts, T. and Yang, U. (1969). *Cytogenetic*. Moscow, Mir

Stack, S.M. and Anderson, L.K. (2001). A model for chromosome structure during the mitotic and meiotic cell cycles. *Chromosome Res.*, 9, 175-198.

Stebbins, G.L. (1971). *Chromosome evolution of higher plants*. Arnold, E. (Ed.), London.

Stevens, A.R. and Prescott D.M. (1971). Reformation of nucleolus-like bodies in the absence of postmitotic RNA synthesis. *Ibid.*, 48(3), 443-454.

Strochkova, L.S. and Kazan'ev, V.V. (1975). Cytochemical peculiarities of a cell cycle. *V. 2. Cytology. Mitotic cell division*. pp. 9-66. VINITI AN SSSR, Moscow.

Stryer, L. (1981). *Biochemistry*, 2nd ed. Freeman, W.H. (Ed.), San Francisco.

Swann, M. (1961). Protoplasmic structure and mitosis. I. Birefringence of the metaphase spindle and asters of the living sea urchin egg. *Exp. Biol.*, 28, 417-433.

Swanson, C.P. and Webster, P.L. (1977). *The cell*, 4th ed. Prentice-Hall Inc., Englewood, Cliffs, Jersey.

Szent-Györgyi, A. (1960). *Introduction to a submolecular biology*. Academic Press, London, New York.

Terasima T. and Tolmach L. (1963). Growth and nucleic acid synthesis in synchronously dividing populations of HeLa cells. *Exp. Cell Res.*, 30(2), 344-362.

The cell cycle. Problems of regulation. (1973). O.I.Epifanova (Ed)., Nauka, Moscow.

Tichonov, A.N. (1999). Molecular motors 2. Molecular basis of biological mobility. *Sorovski educational Journal*, 6, 17-24.

Tokin, B.P. (1987). *General embryology*. 4th ed., Moscow, Visshay shkola.

Tomilin, P.V. (1983). *Genetically stability of cell*. Moscow-Leningrad, Nauka.

van Driel, R., Fransz, P.F. and Verschure P.J. (2003). The aukaryotic genome: a system regulated at different hierarchical levels. *J. Cell Sci.*, 116, 4067-4075.

Vasiliev, Yu. M. (1996). The cell as a miracle of architecture. *Sorovsky Educational Journal*, 2, 36-43.

Volkenshtein, M.V. (1965). *Molecules and life. Introductions to a molecular biophysics*. Nauka, Moscow.

Volkenshtein, M.V. (1983). *Theoretical biophysics*. Nauka, Moscow.

Watson, J.D. (1965). *Molecular biology of the gene*. W.A.Benjamin, Inc. New York, Amsterdam.

Webster J.D. and Vant'Hof J. (1969). Dependence of energy and aerobic metabolism of initiation of DNA synthesis and mitosis in G_1 and G_2 cells. *Exp. Cell Res.*, 55(1), 88-94.

Welch, M.D. (1999). The world according to Arp: Regulation of actin nucleation by the Arp2/3 complex. *Trends Cell Biol.*, 9, 423-427.

Yudin, A.L., Neyfakh, A.A. (1973). Migration of newly-synthesized RNA during mitosis. IV. Content of labeled RNA in nuclei before and after mitosis in *Amoeba proteus.Exp. Cell Res.*, 82(1), 210-214.

Zeuthen, E. (1964). The temperature – induced division synchrony in *Tetrahymena. Synchrony in cell division and growth.* pp. 99-158. N.Y., London.

Zhimulev, I.F. (2003). *General and molecular genetics*. Novosibirsk, Sibirskoe universitetskoe izdatelstvo.

Zholkevich, V.N., Gurvich, A.L., Borisova, T.L., Livanova, T.N. and Shlyakhtina L.S. (1973). On energy discharge prior to cell division. *DAN SSSR*, 209(4), 961-965.

Zosimovskaya, A.I. (1973). Mechanisms controlling the entry of cells into mitotic cycle as studies using specific inchibiters of macromolecular synthesis. *The cell cycle. Problems of regulation*. pp. 104-133. Moscow, Nauka.

Chapter III

Types of Cell Divisions

Abstract

The corpuscular-wave models of the 9 different types of cell reproduction are discussed: mitosis, asymmetric mitosis, activating mitosis, meiosis, activating meiosis, free-nuclear mitosis, endomitosis, cleavage and promitosis. Their comparative analysis was carried out by the following criteria: 1) the state and construction of a biological harmonic oscillator; 2) the number of internal energy sources; 3) the channels of energy utilization; 4) the characteristic changes of the DNA-particle and the DNA-wave during cell division; etc. Particular attention is paid to the spatial disposition of divisible cells and the spindle of their division; the environment that is created inside of a cell structure (or a cell) before the change of the typical scheme of mitosis or one of its variations. The wave nature of such phenomena as a conjugation, crossing-over, apoptosis, etc. was revealed. It is shown that the reproductive cycle of all species of higher plants and animals include the three types of cell divisions, during which the DNA molecules can change or the wave (oogenesis), or corpuscular (male meiosis), or simultaneously corpuscular-wave (female meiosis) its own characteristics.

All enumerated types of cell reproduction are the derivatives from the two universal schemes of a cell cycle, functioning on the basis of: 1) the mechanism of prokaryotic cell division (promitosis) and 2) the mechanism of eukaryotic cell division (mitosis). Mitosis is only one variation, which during evolution originated in the manner of prokaryotic cell division. The others six types of cell divisions arose from mitosis. The standard scheme of a cell cycle with mitosis (chapter 2) was used as a sample for the analysis. The transition from promitosis to the mitosis and from mitosis to the schemes of, accordingly, asymmetric mitosis, activating mitosis, meiosis, free-nuclear mitosis, endomitosis and cleavage are discussed.

The two principally different schemes of the reproduction of cells have been created in the course of evolution: 1) the promitosis or the manner of prokaryotic cell division and 2) the mitosis or the manner of vegetative eukaryotic cell division. The most universal and wide-spread type of cell development is the cell cycle with mitosis. Today in addition to mitosis is described a number of other types of cell divisions - asymmetric (or differentiating) mitosis, meiosis, free-nucleus mitosis, endomitosis and cleavage. All of them are different variations of mitosis and are derived from it. For their analysis, we will use the corpuscular-wave scheme of the cell cycle of eukaryotic proliferous cell, as a model (chapter 2). For each type of cell division, we have tried to find the moment at which a change takes place in the scheme of mitosis or on one of its variants. The emphasis has been on the physical state of the cell or complex cells and the physical method that is caused by the corresponding transition. For each type of cell division we will analyze, mainly, the state of DNA-particles, DNA-waves, the presence or absence of second (2nd) internal source, and the channels of energy movement into the cell.

Among the analyzed types of cell divisions only the promitosis (the manner of the division of prokaryotic cells) occupies a special place. Its construction occurred before mitosis and served as the basis for a more perfect mechanism of cell reproduction - mitosis. Prokaryotic cells are as a whole weakly differentiated and less specialized. Nevertheless, the universal mechanism of replication, transcription and translation, and main energy channels, which differ by the type and character of used energy - centrifugal (on ATP-ADP base) and centripetal (on processing base) energy streams are inherent to them already.

In this chapter a 9 corpuscle-wave model of the types of cell divisions of prokaryotic and eukaryotic cells with an analysis of reason for changes in the scheme of mitosis are discussed.

Mitosis

The detailed scheme of plant cell development was given in the previous chapter. Its cell cycle includes two periods – evolutionary and involutionary; and a four-quarter cycle – latent development, mitotic development, cell birth and ontogeny. The cell development takes place in an open system and a closed system. Energetics is provided by four in depend of sources: the two internal – the energy of the biological harmonic oscillator and the metaphase chromosomes, and the two outer sources - the energy of photosynthesis and respiration. The processes of replication, transcription and translation are separated by cell cycle stages and used for this purpose the energy from the different sources the "DNA-particle - DNA-wave" system changes in every quarter cycle. Below is a short biophysical characteristic of the cell cycle, taking into account the quantum mechanical nature of cells.

We present the mechanism of mitosis in the form of a mechanical system consisting of two coupled pendulums (Figure 11-a). After a doubling the DNA molecule consists of two DNA-particles and one DNA-wave. The length and weight of pendulums are equal because we are concerned with identical DNA-particles. The presence of springs says that the two DNA- particles are still connected by chemical bonds. The pendulums oscillate with the same frequency, i.e., have the same energy. Assume the energy of a cell system (after DNA-wave radiation) corresponds to the second energy level (2hω). Now the spring breaks (dividing the doubling DNA-particle) and carefully separates the pendulums in different directions, then the oscillation frequency of each of them (two DNA-waves) does not change and will be equal to: $E_1 = E_2 = 2h\omega$.

Thus at the end of mitosis, each daughter cell receives the same DNA-particle (a hereditary basis of the cell) and the same DNA-wave (an energy basis of the cell). Regarding the mother cell, it is identical with the daughter cells on a hereditary basis (by DNA-particle) and differs from them on one level of energy basis (by DNA-wave).

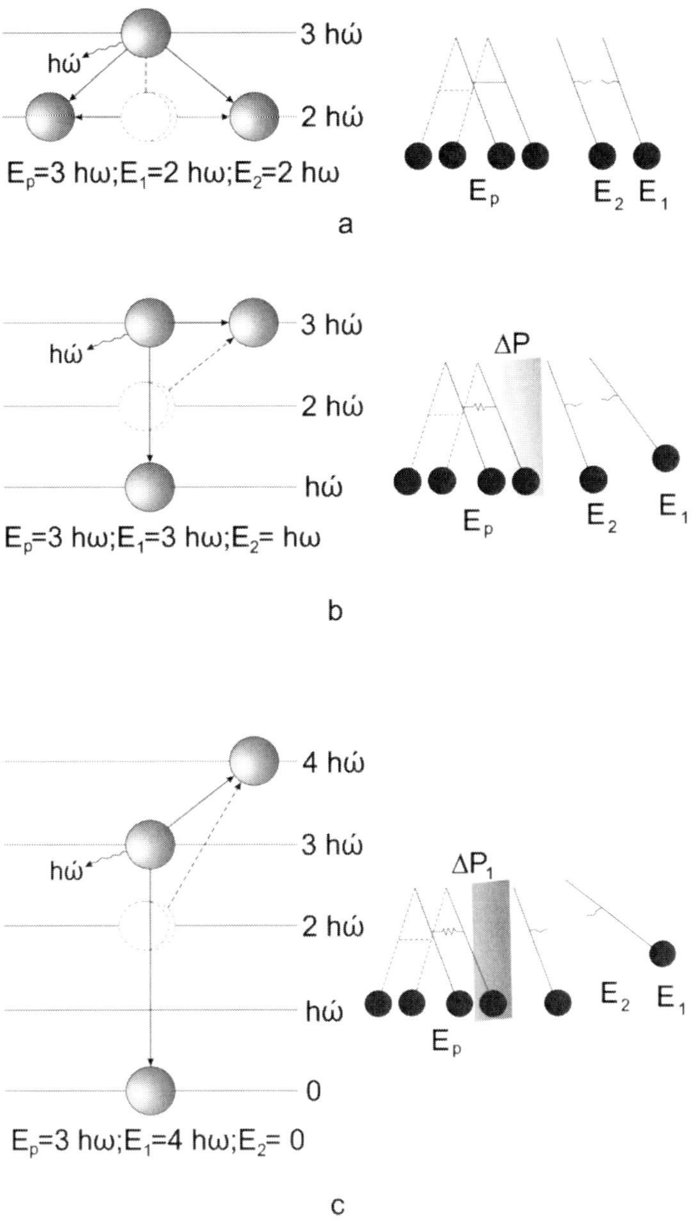

Figure 11. The corpuscular-wave scheme of cell division by type: a) mitosis; b) asymmetric mitosis; and c) activating mitosis.

Asymmetric Mitosis

Beginning with Aristotle to the present day, scientists have discussed such questions as: the reasons for the transition to sexual reproduction, the problems of embryonic development and differentiation of organisms, an increase of the ordered complexity of biological systems during ontogeny and the disordering of them during sexual reproduction and the nature of resuscitation and loss of totipotency. According to our understanding, the answer on some questions is in the wave nature of the genome, as well as the existence of such cell divisions as an asymmetric and activating mitosis. The corpuscular-wave models of these types of cell divisions we will discuss below.

Asymmetric or differentiative mitosis is at the end of the formation of daughter cells that have a different fate. One cell remains in the meristem, while the other becomes a part of the plant body, and its development follows by the way of differentiation (Mazia, 1961; Ivanovskaya, 1983; Knoblich, 2001; Gavrilova, and Rudakova, 2003; Gilbert, 2010). For the first time, this type of mitosis was found in bacteria (Lawler and Brun, 2006) and is widespread among animals and plants (Fowler and Quatrano, 1995; Stephens et al., 1995; Gallagher and Smith, 1996; Lin and Schagat, 1997; Roegiers and Jan, 2004; etc.). According to literature data (Nelson, 1992; Roegiers and Jun, 2004; Terskikh et al., 2007, 2009) the cell, for entry into asymmetric mitosis must satisfy a number of requirements: 1) the definite polarization of mother cells; 2) the mitotic spindle must be positioned along the axis of polarization and 3) the daughter cells must contain different determinants that are defined their fates. In higher plants the cells are divided by the principle of the cells of the apical meristem of stem and shoot. A quality of apical growth is that the nucleus during divisions moves along the growth axis (Gavrilova and Rudakova, 2003), and the cells are polarized along the main axis of the plant.

We think that the "determinant" is the energy state of the biological harmonic oscillator of two daughter cells. In this case the change of mitosis scheme could take place as follows. At some moment, one of meristematic cells (which occupy an apical position in a multicellular organism) turned so that its polar axis and spindle division began to coincide with the direction of gravity. At present, the role of gravity and polarity in the development and evolution of biological systems has been confirmed many times (Dure, 1976; Ivanovskaya, 1983; Merkis, 1990; Fowler and Quatrano, 1995; Stephens et al., 1995; Drubin and Nelson, 1996; Tairbecov et al., 1997). The density of the cytoplasm along the polar axis is identified by the effect of gravitational forces on the cell. Since each intracellular lower layer must support the weight of the upper layer, then the density of the cytoplasm must increase progressively from the apical to the basal cell pole.

The movement of biological energy (i.e., the energy from the 1st inner source) during asymmetric mitosis is show on the system of coupled pendulums with an initial energy level of 3hω, and after the DNA-wave radiation of 2hω (Figure 11b). For this purpose we divide a mother cell at the horizontal sections along the apical-basal axis. Since the cell is polarized and divided transversely, the 2nd pendulum during division is moved into a region of dense cytoplasm, which slows down its movement. During separation it gets a blow from the coupled 1st pendulum. At that moment the 1st pendulum obtains a quantum of energy (E_1= 2hω + hω), and on the 2nd pendulum, this quantum of energy is lost (E_2= 2hω-hω). If at this moment the pendulums are separated they will have a different energy, which is equal to, accordingly, 3hω and hω. Apparently, during asymmetric mitosis the transfer of biological energy from one (basal) chromosome to another (apical) chromosome takes place.

The cell morphology is directly dependent on energy from components of its genome. The ultrastructure of the meristematic and differentiating cells is always distinguished after asymmetric mitosis (Clowes, 1971; Gamaley, 1972; Danilova and Kashina, 1980; Gilbert, 2010). In figures 12-a and 12-b is shown the cell fragment from the apical meristem of root. The nucleus is round, with a large nucleolus. The cytoplasm has slight vacuolization and the vacuoles are small, often with osmiophil inclusions. The cytoplasm contains many ribosomes. The numerous mitochondria are round or oval. The plastids are presented by typical leucoplasts with a dense stroma and a poorly developed internal structure. In these cells there are many lipid droplets.

The derivative or differentiated cell, which has lost its energy during division and has started by way of differentiation, is characterized by a slightly different ultrastructure (Figure 12-b). The main distinguishing sign is the intensification of vacuolization. Small vacuoles increase in volume and coalesce, but the central vacuole is still out. The ultrastructure of the nucleus, mitochondria and plastids do not change significantly. Most likely, at the further transition of the DNA-wave to the lower energy levels will decrease the intensity of the biofield, the forces of which are, for the most part, arranged in the ultrastructure of aging cells.

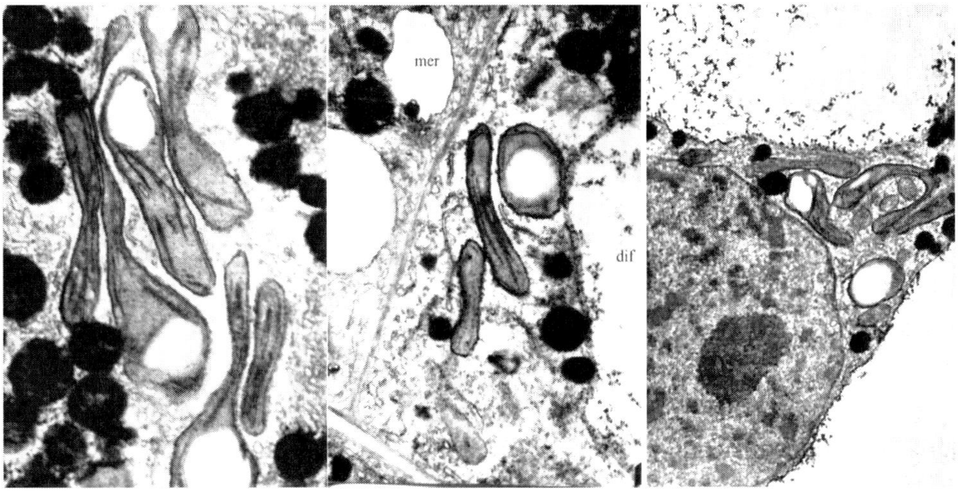

a b c

Figure 12. The cells of the root of *Pinus sylvestris* L.: a) meristematic cell; b) meristematic and differentiating cells after asymmetric mitosis; c) differentiating cell; mer – meristematic cell, dif - differentiating cells.

Thus, the asymmetric mitosis is connected with the change of genome wave characteristics only. It is finished by the formation of one daughter cell, a genome of which is an exact copy of the genome of the mother cell according to the state of the DNA-particle and the DNA-wave before the radiation of the DNA-wave occurs (3hω). For the new division this cell again becomes the mother cell with the energy potential of the meristematic cell. The second daughter cell (i.e., the derivative cell, which lost the energy during division), in accordance with the second thermodynamic law, has descended down two levels (hω), and as the cell of plant body, is accomplishes this by way of differentiation. And it will take place every division by means of asymmetric mitosis. That is why theoretically the effectiveness of meristematic tissue on cell division is endless during asymmetric mitosis. Probably, differentiation and asymmetric mitosis are considered as the first step towards to the appearance of multicellular organisms.

Activating Mitosis

Activating mitosis occurs only in the generative cell and participates in the construction of all reproductive cell cycles of animals and plants that reproduce sexually. We call this type of cell division activating mitosis, because during these divisions the biological harmonic oscillator is filled up by energy and the generative cells enter into a new cycle of ontogenesis (Kuznetsova, 2002, 2003). It originated in the two daughter cells with a diametrically opposite fate. One of them that has received a specific amount of biological energy and is raised to an energy level of $2h\omega$, $3h\omega$, and above, "is rejuvenated" during this process and is passed as in the reproductive process (megaspore, egg) or starts a new cycle of ontogeny (zygote). The second daughter cell which lost the energy will have died (synergids, prothallial cells). In higher plants, a typical example is the stage of oogenesis with two consecutive activating mitoses. The archegonium initial cell is divided twice by means of activating mitosis and creates the 4 different cells. One of them - the egg (the cell of apical pole) after fertilization will give rise to new reproductive structure – the zygote. All the other daughter cells (the primary neck cell and its derivatives and the ventral canal cells – the cells of the basal pole) quickly die and go out of the reproductive process. During activating mitosis as during asymmetric mitosis changes only occur to the wave properties of cell genome and stay constant with the corpuscular ones. These two types of mitosis are close to each other by the main coordination requirements– the spatial relationship of the polar axis of the mother cell and the spindle of division with the direction of gravity. However, we have distinguished the activating mitosis in an in depend of type of cell division, because for existence it requires the supplementary physical conditions the description of which will be discussed below.

The analysis of two mechanisms "filling" with energy from the biological harmonic oscillations during activating mitosis will be shown in the examples of two reproductive stages – the megasporangiogenesis and microgametophytogenesis of conifer plants (Singh, 1978; Kozubov et al., 1982; Kuznetsova, 2002).

1. The reproductive cycle of all species of conifers begins from the formation of the primordium of the megasporophyll. At this time the primordium represents an area located near the circumference and is closely pressed to other meristematic cells. Accordingly, their polar axes are at different angles towards the direction of gravity. The initially cell of ovule becomes one (or a few) cells of an apical pole from the surface layer of the primordium. Because the polar axis of an initially cell of ovule coincides with the direction of gravity this cell is a polarized structure. An initial undergoes two axial transverse divisions, or activating mitosis. During these divisions the karyokinetic spindle has the same direction as the initial polar axis (Kuznetsova, 2002). After the 1st division the cell of a basal pole occupies the place of the initial ovule. The cell of the apical pole is pushed inside the primordial tissue. During this movement the apical cells move apart (often destroying) the primordial cells, for which it needs a large amount of force.

2. Gametogenesis (microgametophytogenesis) is another stage at which the DNA-wave of a microspore is transferred to higher energy levels. In gymnosperms the pollen grains are multi-celled (Sterling, 1963; Singh, 1978; Kozubov et al. 1982). In Scots pine the microspore undergo five activating divisions and form two male gametes. The larger sperm takes part in the fertilization, and the second one dies. All cells of the basal pole (two prothallial cells, the cell tube and the sterile cell) are dying as well. The pollen grains have a firm outside envelope

or a sporoderm, which is not capable of stretching. I.e., the first division of microspores changes into a fixed volume. Its division is initiated by a DNA-wave radiation and the quantum of free energy enters into the cell. A part of this energy is turn into a kinetic energy. This causes an increase of pressure in the pollen grains and the transfer of their DNA-wave to higher energy levels.

From the above mentioned, it follows, that in the reproductive cycle of all species of plants and animals there are special cell divisions, a reproductive stage, a morphological structure and a physical mechanism for the DNA-wave transference to higher wave energy levels. In higher orders of plants there are two such methods: 1) the creating of a density gradient between the cytoplasmic layers in a body of generative cells; and 2) the increase of intracellular pressure in a single generative cell (microspore) or complex cells (female gametophyte). The "charge" of the biological harmonic oscillator by energy during sexual reproduction is a strictly genetically determined process, which is carried out by transfer of biological energy from the chromosomes of basal pole to the chromosomes of apical pole. In higher order plants this reproductive stage, which we call an activation stage, usually consists of two or more activating mitoses. In the species reproductive cycle the number of given stages and it construction is also under genetic control.

An analysis of the forces that act on the apical pole of the initially cell of ovule is now examined. Their dependency on the direction of action of forces into the system can be divided into two groups. On the one hand - this is the kinetic energy of the molecules of cytoplasm, which is caused by cell growth and advances into the primordium. In the same way they are acted upon by gravity. On the other hand this is a resistance force, which is proved by the characteristics of the primordial cells. These forces are directed towards each other and along the polar axis of an apical cell. Therefore the density gradient between the cytoplasmic layers is much more in the initially cell of ovule than in the mother cell during asymmetric mitosis.

The dualistic scheme of an activating mitosis and the energy transfer from one chromosome to another ones is well observed on the same system of coupled pendulums with an initial energy of 3hω, and after the radiation wave-DNA of 2hω (Figure 11-c). By means of separation the 2nd pendulum enters into the layer of rapidly condensing cytoplasm, which causes it to stop. As the pendulums are connected, it receives from another pendulum a strong blow during separation. During this blow the 2nd pendulum gets two quantums of energy. The spring breaks at this moment. The energy of the 1st pendulum after their divergence is equal to 0 hω (E_1 = 2hω-2hω), and the 2nd one is equal 4hω (E_2 = 2hω +2hω). I.e., after division the energy state of one daughter cell becomes higher on one level (4hω), than in mother cell before radiation (3hω), the other daughter cell, on the contrary, decreases by three levels (0 hω). The cell, which received the energy initially, begins a new cycle of ontogenesis. The other daughter cell which lost the energy, now dies. If the mother cell begins the activating mitosis from a higher level (4 hω, 5 hω and above), then by conservation of energy, at a lower level, its biological harmonic oscillator is able pass through a larger number of levels, and the difference between the energy state of the apical and basal generative cells would be greater.

The possibility of the transference of biological energy from cell to cell during activating mitosis explains the nature of apoptosis as well. Apoptosis, or programmed cell death is the phenomenon, which is frequently occurring in plants and animals during sexual reproduction and tissue differentiation (Greenberg, 1996; O'Brien et al., 1998; Galitsky, 2005). At the present time there are many hypotheses for explaining the nature of apoptosis. Our understanding about the physical nature of apoptosis has widened after the unification of the experimental data with the laws of energy conservation and the wave nature of the genome. The "charge" of the genome of future plants by biological energy takes place during the specific number of activating mitosis in the one part of the generative cells at the expense of the death of another part of the generative cells. It follows that the apoptotic cell is a generative cell, which has lost its own biological energy, and for this reason its death is guaranteed.

Thus, during activating mitosis the two daughter cells are formed identical with the mother cell on a basis of the DNA-particle, and are distinguished on a basis of the DNA-wave. After division the apical cell, which has restored its energy to a species-specific level, passes into ontogenesis or the next stage of the reproductive cycle (Kuznetsova, 2002, 2003). The basal cell, which has lost energy, is developed by means of deep differentiation and then dies. The transfer of biological energy is carried out from chromosome to chromosome where, in generative cells, are created two type of specific physical conditions (an increase of intracellular pressure or the creation of a density gradient between the cytoplasmic layers). Between the asymmetric mitosis and the activating mitosis are both similarities and differences. A common characteristic is the physical mechanism of the transference of biological energy from chromosome to chromosome and the coincidence of the spatial organization of mitosis. However for activating mitosis realization there are required supplementary conditions, such as, the special reproductive structures of vegetative (primordium) or generative (female gametophyte) cells and the change of intracellular pressure.

Meiosis

The two types of cell divisions – meiosis and activating mitosis are the obligatory components of all reproductive cycles of plants and animals that reproduce sexually. Moreover, we are confident that the organism transition on a multicellular level became possible only after the origination of these two variations of mitosis: meiosis for a change of the DNA-particle and activating mitosis for a change of the DNA-wave. The cytology of meiosis is now carefully studied and described in many research works (Darlington, 1931, 1965; Rhoades, 1961; Johri, and Lewis, 1965; Swanson and Young, 1965; Sounson, 1969; Cytology and …, 1975; Sybenga, 1975; Swanson and Webster, 1977; Chentsov, 1985; 2004; John, 1990; Bogdanov, 1991; Zavarzin et al., 1992; Alberts et al., 2002; etc.). The distinguishing trait is that after the doubling of the DNA-particle follows the two divisions of the nucleus, giving rise to a four haploid gamete. There are the two types of meiosis - an ordinary mitosis, which forms four viable gametes (male meiosis) and an activating meiosis, where the number of gametes is reduced to 2 or 1 (female meiosis). The more ancient type of meiosis is the male meiosis. Analysis of the corpuscular-wave model of male meiosis will begin with a description from a norm position or the state of the chromosome-particles and chromosome-waves in mitotic cells after DNA replication.

During the interphase the spatial disposition of the chromosomes-particles inside the nucleus is always fixed (Monachova, 1990, 2004; Chentsov, 2004). Their set resembles a truncated cone, which lies across the force of gravity (Figure 13-a). Each chromosome on the surface of the cone has its own niche and is connected by a nucleus envelope at three points – in a centromere and two telomere parts. Homologous chromosomes are territorial close to each other (Monachova, 1990, 2004). Centromeres are evenly distributed around the circumference on the upper base of the cone. The telomere ends of the chromosomes are on the bottom base of the cone and linked end to end. In this case it joins the telomere ends by means of the lines, and we acquire a system of triangles which lie in different planes and evenly cover a cone on the outside.

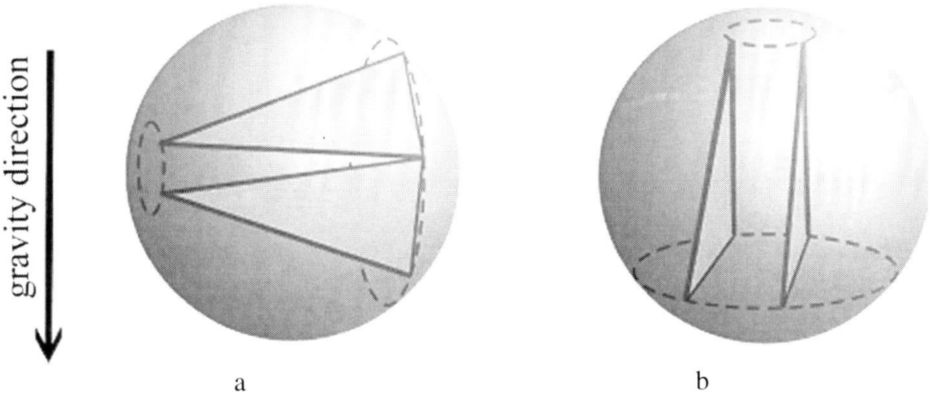

Figure 13. The spatial disposition of chromosome particles inside the nucleus during a beginning (a) and end (b) of meiotic prophase.

Analysis of the physical model of the reasons for the beginning of meiosis will follow on the example of the life cycle of *Bacillariophyta* (Komarnitskii, 1975; Life of plants, 1978; Reyvn et al., 1990; Mann, 1993; Lee, 2008). All species diatoms are diploid organisms. On the outside, their volume is fixed strictly by the flint armor. It consists of two halves – an epiteca (the greater external part) and a hypotheca (the lesser inside part), which is not capable of stretching. A vegetative reproduction of all diatoms is peculiar because of the presence of flint armor. After division the daughter cells receive half of mother envelope and begin to restore the missing part of the envelope, it is known that the only inside part is a hypotheca. Therefore, it will form daughter cells of different sizes. The cell which received the epiteca preserves the size of mother cell and the second cell which received the hypotheca becomes smaller. Therefore the vegetative reproduction of diatoms is accompanied by a decrease of individuals, which can reach up to 2 times or more. Restoration of cell size takes place during sexual reproduction. Meiosis begins when the cells have reached the specific minimum size during frequent mitotic divisions. Let us consider what physical conditions are created during the progressive decrease of the cell size of individuals during ontogenesis, which are necessary for switching from mitosis to meiosis.

It is known that the behavior of all molecules, which are distributed in a limited volume of solvent, obey the gas laws. Then the behavior of the biomolecules, which "dissolved" in a cell protoplast, the volume of which is rigidly fixed, should obey the gas laws as well. The equation of the state of ideal gas is related to pressure, temperature and volume by means of the ratio:

$$\boldsymbol{PV = RT} \qquad (1)$$

wherein

D – density, m – mass and N – number of molecules.

The diffuse phase (cell sap) of intact cells is a reversible colloid system, which is able to turn into a solid from the gel and back again. It is considered that the many changes in behavior, movement and cell division of cells are associated with viscosity (Heilbrunn and Wilson, 1948; Heilbrunn, 1956). It has been proven that moderate pressure increases the cell viscosity. This occurs during mitosis as well. This is a rather labile system, which depends on the action of internal and external factors (temperature, pressure, biofield etc. state).

The notion of viscosity (or internal fluid friction) and the density of matter are directly related. The density of a substance is derived from the formula:

$$D = \frac{Nm}{V} \tag{2}$$

wherein

P – pressure, V – volume, T – temperature, R – universal gas constant.

Nm – the number of biomolecules on their mass is a constant quantity, because all synthetic processes in a cell are under strict genetic control. It was experimentally proven that the amount of protein, RNA, lipids, etc. are always doubled before each cell division (Zosimovskaya, 1973; Lehninger, 1985; Alberts et al., 2002). This signifies that at the end of G_2 period all generations of diatoms contain the same number of biomolecules. However, their concentration in the more small cells is higher. If we proceed from the information gathered by formula (1) and (2), the concentration increase leads to two important consequences: an increase of intracellular pressure and viscosity. All the more so, both factors – a pressure and a biomolecule numbers – are created in the increasing of cell viscosity.

We have all the reasons for the assumption that pressure and viscosity are the main physical factors by means of which is initiated a meiosis in other taxonomic groups of plants as well. In species of *Conjugatophyceae, Volvocophyceae, Ulothrichophyceae,* etc. with a haplobiont life cycle the zygote is only a diploid cell. From the outside it is covered by a stable and dense envelope. After the dormancy period, the first zygote division is meiosis for the following reason. As mentioned above, DNA-wave radiation and an entering into the cell of the quantum of free energy is preceded by any type of cell division. This energy is spent through a number of channels, including cell growth as the volume of a zygote is strictly fixed and is what leads to an increase of intracellular pressure and the beginning of meiosis.

In all higher order plants the development of archegonial cells and microsporocytes are undergone in the central part of the microsporangium (Poddubnaya-Arnoldi, 1976; Embryology of ..., 1984), where the physical conditions (a higher level of intracellular pressure and viscosity) which are necessary for the start of meiosis are created as well. In Figure 14 it can be seen that close to a sporogenous tissue of the arhegonium of *Ouercus robur* is the powerful layer of the microsporangium wall (endothecium, exothecium, middle layer and tapetum), which will carry out the function of a outer "capsule". Therefore, the intense dividing and growing generative cells try a strong resistance from the anther tissues. The end result of these two divergent processes is the transition of the mother cells of microspores from mitotic to meiotic divisions.

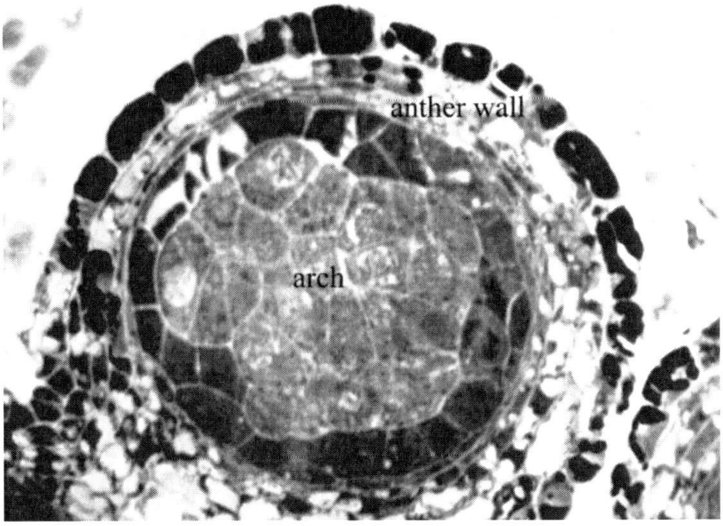

Figure 14. Another wall and sporogenous tissue of a microsporangium *Ouercus robur;* arh – sporogenous tissue of arhegonium anther wall – layers of microsporangium wall.

The increasing of intracellular pressure and viscosity will make a number of alterations in the standard scheme of mitosis. Our task now will be to analyze a cell cycle under the new physical conditions and determine at what moment the changes begin on the standard scheme of its development. As known, meiosis in the same manner as mitosis is divided into a number of successive stages. Today the evolutionary conservatism and the universality of all meiotic stages and meiosis as a whole are confirmed in different eukaryotic organisms (John, 1996), though their interpretation in this chapter is essentially different from current cell theory. Before the beginning of the description of meiosis, we decided once again to say, what is necessary to remember is that there are many biological phenomenon that are connected with the wave nature of DNA and the quantum-mechanical nature of a cell and how there are many "dark holes" in this field of our knowledge.

The cell cycle of meiosis begins from the DNA-wave radiation as well. Its energy initiates a DNA-particle doubling and a centriole formation, which sets in motion the mechanism of chromosome coiling. If friction is negligible, the biological harmonic oscillator is characterized by a high quality factor (Q). From the above mentioned, the viscosity of cells becomes higher during the diatoms ontogenesis. Therefore the mechanism of meiosis will be realized in the new physical conditions. Firstly, the friction of nucleoplasm begins to increasingly absorb the energy of the biological harmonic oscillator and, consequently, diminishes the amplitude of the DNA-wave oscillation. The more viscous nucleoplasm will cause the exponential extinguishing of the DNA-wave. Secondly, when the amplitude of the DNA-wave oscillation is decreasing the probability of simultaneously entering into the area of coiling of 4 isomers becomes greater (see Chapter 2). This decreases the speed of chromosome coiling and the duration of the G_2 period become considerably longer. This is confirmed by a weak chromosome coiling which has the form of thin threads during leptotene (Johri and Lewis, 1965; Sounson et al., 1969; Cytology and ..., 1975). Thirdly, there is the pressure interval when the curve of the intensity of the DNA-wave radiation drops sharply. At this time the interval between two successive radiations is greatly increased. The time interval between two successive radiations is then prolonged and the duration of the cell cycle on a whole is increased. It has been shown that the meiotic divisions when compared to mitotic divisions, continues for a longer period of time in all plants and animals (Rhoades, 1961; Johri, and Lewis, 1965; Cytology and ..., 1975). It should be noted that the increase of the duration of G_2 period and the cell cycle is a necessary but not sufficient requirement for the cell transition from mitosis to meiosis.

The rotation of centrioles in a more viscous medium sets in motion the chromosomes. They begin to intertwine around themselves and remain associated by means of centromeric and telomeric ends with a nucleic envelope. Experimental studies have shown that the twisting of chromosomes occurs during the early meiotic prophase (Raven, 1961; Cytology and ..., 1975). An unusually twisted chromosome in a proleptotena even got a special name – a prochromosomes. The chromosomes are twisted until the returning force reaches its maximum. After that they quickly unwind and lose contact with the nucleus envelope. Before that moment, the chromosome set represents the physical body in the form of a truncated cone, which is relative to the direction of the gravity and is situated in a cross-section (Figure 14-a). This physical body undergoes the effect of torque, which is produced by gravity. That's why in the moment the chromosome separation from the nucleus the cone is rotated inside the nucleus and "stands" on the lower base (Figure 14-b). The change of the spacial orientation of the chromosomes during a leptoneme is confirmed by the appearance of figures of a so-called "bouquet" in which the chromosomes are arranged along in the direction of the gravity (Raven, 1961).

The leading event, which is necessary for a transition to the meiotic scheme division, takes place during a chromosomes turning. The condensation chromosomes after separating from the nucleus envelope begin uncoiling. At the same time as the lifting and uncoiling chromosomes will be the flow of torsion waves of along their shoulders. The waves will set in motion the chromosome shoulders, turning each chromosome in the nucleus so that the shoulders of the homologous chromosomes become one with the parallel disposition. If the chromosome set is imagined as a system of triangles, the sides of the triangles (shoulders of homologous chromosomes) will be parallel (Figure 14-b). After a turning the uncoiling and elongated chromosomes soon reach the nucleus. At the moment of impact to its envelope, the DNA-wave and the process of chromosome coiling are resumed again. It seems obvious that the turn of chromosomes and the change of their spatial orientation are the main events during leptotena. After this stage the passage from mitosis to meiosis is an irreversible process.

However, it should be noted, that at the beginning of meiosis it is impossible to define a definite size or number of generative cells, which enter to the meiotic division. In the cell generations only raises the likelihood of its realization. That is why diatoms (starting from the specific cell volume) the meiosis are detected in the cells of different size and generations (Komarnitskii et al., 1962). The same pattern was found in salamanders (Gurvich, 1977). From the primary spermatogenous cell at the same time is entering into meiotic divisions of spermatocytes of the 8th, 7th and 6th generations.

The process of the condensation of chromosomes continues when a DNA-wave is restored. But now, on the mechanism of coiling influences is gravitation. Its forces make isomers entering into the zone of coiling a more orderly process (see Chapter 2). Therefore the rotation of the isomeric links does not occur randomly, as during mitosis. This process takes place in the local areas of the chromosomes - chromomeres, when each new coiling fragment "is attached" to the previous one. It is shown that the chromomeres are formed during coiling and a chromomeres stretching can get a thread of equal length (Rhoades, 1961; Sounson et al., 1969). The same with the oscillation frequency of a DNA-wave and the DNA-particle structure of the same type are responsible for the species-specific size, number and arrangement of chromomeres on the chromosomes (Swanson and Webster, 1977; Kiknadze, 1972). The process of chromosome coiling and the decreasing of their length continue during the entire G_2 period, which is greatly extended. During the leptotena, zygotena and pachytena stages the chromomeres gradually become larger and their number decreases. The cell during the process of chromosome condensation will be drawing close to the moment of irreversibility, when the destruction of the DNA-wave and the mechanisms of the meiotic division are realized.

Zygotena is, the stage of active conjugation of chromosomes, conjugate the homologous chromosomes which the draw together during leptoneme. Before a convergence the chromomeres of one homologue are exactly opposite of the chromomeres of other homologue. The zygotena finish by the formation of specific chromosomal ensembles (bivalents), which are represented as a pair junction of doubled homologous chromosomes.

During zygotena a DNA molecule still exists in the form of a particle and a wave simultaneously. The nature of the conjugation phenomenon is completely determined by the corpuscular and wave properties of the chromosomes. Homologous chromosomes are the same by the shoulders length, the angle between them and the number and disposition of the chromomeres (Rhoades, 1961; Swanson and Young, 1965; Alberts et al., 2002; Zhimulev, 2003, etc.). Therefore, they can be considered as identical particles. One of the important consequences of quantum mechanics is that the waves of identical particles, which find themselves in the same place, must interfere with each other. The end result of the interference of two homologous chromosome-waves, is the formation of one resulting wave or one bivalent-wave. The start of the conjugation processes helps the new position of the chromosome set in a nucleus and where the chromosome-particle is firmly connected with the nuclear membrane at three points (in centromere and two telomere parts). The conjugation process or the formation of a resulting bivalent-wave between homologues may be realized in different ways: from the polar ends, one or several points of the chromosome-particles. Then, this process gradually spreads over the whole chromosome. As a consequence the homologues in the bivalents become indistinguishable in the zygotena end.

The next stage is pachytena or the stage of the thick threads for the following reason. Apparently, as a result of the wave interference on the identical particles (if the phase difference is $- a_1 - \alpha_2 = 0$), there is an increase in the intensity of the resulting wave. The intensity of the resulting wave is higher than the sum of the intensities of its wave components. That is why after a conjugation appears the thick pachytena chromosomes, a number of which are haploid.

The leading event of the pachytena stage is a crossing-over. According to the laws of quantum physics, the waves of closely related complex objects during this interference always have a prior probability for an internal exchange between their components. This explains the physical nature of the crossing-over phenomenon. Undoubtedly, the chromosomes are a complex object. That is why after formation of the resulting bivalent-wave the nucleotides of homologous chromosomes periodically exchange their places. The revelation of a synaptonemal complex during pachytena is the experimental confirmation of the presence of a nucleotide exchange between homologous chromosomes (Roth and Ito, 1967; Westengard and Wettestein, 1972; Page and Hawley, 2004). In the M. Westengard and

D. Wettestein (1966, 1970, 1972) works it is shown that the homologous chromosomes of *Neottiella rutilans* are associated with the nuclear membrane during pachytena. A synaptonemal complex is observed along the whole length of the chromosomes. The processes of genetic recombination which take place during the crossing-over, give rise to the four new DNA-particles, whose chemical formulas are not the same as the initial ones.

It is known that all vegetative cells receive the same DNA-particle during mitosis. Therefore, the appearance of conjugation and crossing-over by means of which changes a DNA-particle and expands a genetic variability of species is an important step in the evolution of the sexual process. In Protozoa, which are characterized by the imperfect mechanism of meiosis, the conjugation of homologous chromosomes takes place late (Cleveland, 1949). It is not prolonged and not very close. In plants and animals a conjugation chromosome is a more perfect process. After the conjugation process is over a nucleus has the appearance as if it contains a haploid set of chromosomes (Kushev, 1971; Cytology and ..., 1975; Zavarzin et al., 1992; Alberts et al., 2002). This shows how the mechanism of meiosis was developing during evolution.

During the transition from a pachytena to zygotena the majority of synaptonemal complexes disappear. During diplonema (a stage of double threads) a split of doubled chromosomes on the chromatids and the separation of homologous chromosomes from each other take place. Perhaps, the chromosomes are divided on the chromatids at the moment of the flying away of the centrioles, which is the result of the DNA-wave destruction. The data of cytological study is showed that during mitosis prophase and diplonema stages are functioning with similar processes. But in the first case the doubled chromosomes are divided (2 DNA-particles), and in the second the bivalents are divided (4 DNA-particle). Let us analyze the stage of diplonema in more detail.

It should be noted that during meiosis the process of chromosome condensation by means of internal coiling and the formation of complex isomers, is happening in the same manner as during mitosis. During meiosis it is included in the stages of G_2 period, leptotena, zygotena and pachytena. The destruction of wave function on DNA occurs at the moment when the length of coiling bivalent-particles becomes smaller than the diameter of the nucleus. When it disappears the destruction of the biofield occurs, by means of which the two positively charged centrioles far into the inner region of the nucleus. And the centrioles fly away in opposite directions. After the 1st period of the coiling and splitting of chromosomes the chromatids represent the single piles of complex isomers with two vertical chains of polarized charges – a positive and a negative. Therefore the centrioles during a flying away are taken away with the chromatids by taking one each from the homologous chromosomes. In this time the two pairs of homologous chromatids stretch in different directions (Hugher-Schrader, 1943). The division is going by the semi-conservatively way of reproduction. The bivalents after the destruction of the resulting wave break up on their homologous components as well.

Between the two centrioles, as between the two positively charged particles is formed an electrical field. The chromatids of bivalents begin to do the rotational movements. As mentioned above, the two chromatids of the homologous chromosomes after splitting were close to each other. Therefore during the turning they are attracted to the segments of chromatids with opposite charges and are pushed off from the segments with similar charges. The chiasm is a structure, which forms during the diplonema for the attraction and keeping of the homologous chromatids in a bundle. Sometimes between homologous chromatids there is an absence of chiasms. In this case, they form the two univalents. During metaphase these univalents are not include at the metaphase plate, and in depend ofly of each other they travel away to the poles (to different poles or to the same one) (Petrov, 1973). The number of chiasms may be different. According of C. Darlington (1965), the number of chiasms is proportional to the length of chromosomes: from a single chiasm at the bivalents of short chromosomes to 5-6 chiasms at the bivalents of long chromosomes. The bivalents depending on their number can take different shapes – a cross, a ring or a loop (Petrov, 1973; Swanson and Webster, 1977; John, 1996).

During diplonema the bivalents continue the process of coiling (a condensation with internal coiling) and rapidly shorten. Then each chromatid is coiling independently from the other ones by means of the external rotation of complex isomers (a condensation with external coiling). This leads to the neutralization of the charges and the shoulders of the chromosomes gradually lose the dipole structure. After the neutralization of charges the chromatids are pushed away from each other. According experimental data the chromatids cannot get entangled between themselves. In our model it is explained by the quantum nature of homologous chromatids (see Chapter 5). The number of chiasms decreases by means of their terminalization as far as the advancement of the process of chromosomes coiling (diakinesis stage). During diakinesis the nucleolus disappears, the nuclear membrane breaks down and the bivalents go out near the region of the equatorial plane.

During metaphase I the bivalents are always shorter and thicker than mitotic metaphase chromosomes (Darlington, 1965). This is verified by the quantity of energy held in them. At this time the centromeres in the bivalents (as the two negative charged particles) lie at an equal distance from the equatorial plate, forming the two narrow rings with half of centromeres in each one (Swanson and Webster, 1977; John, 1996). During anaphase I the homologous chromosomes are moving to different poles. When the centromeres come nearer to the centrioles their charges are neutralized. The centromere ring is narrow, therefore it at a blow it divides the centriole or the molecule of polymeric actin into two parts. They, as the two positively charged structures are moving away to opposite sides. Thus the two new centrioles or cell centers originate, which will provide the mechanism for the second meiotic division, when the homologous chromatids are separated. The change of the DNA-particle and the DNA-wave during meiosis is shown in Figure 15.

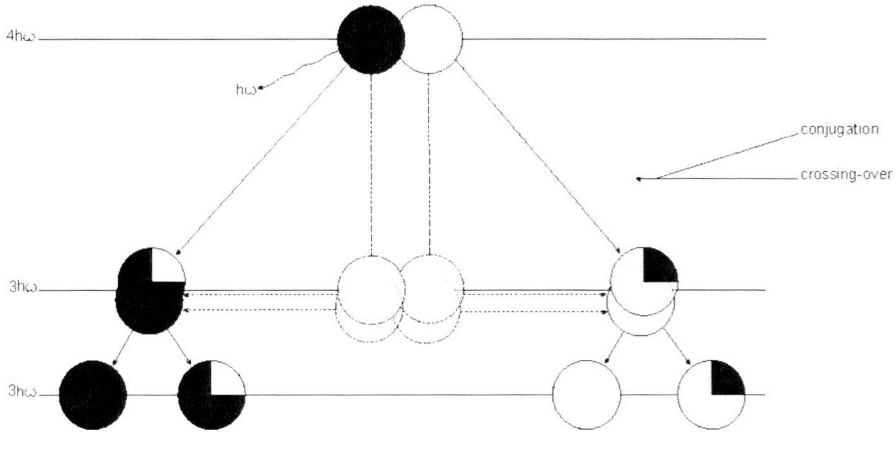

Figure 15. The scheme of the change of DNA-particle and DNA-wave during male meiosis.

Thus, the meiosis is a mechanism for changing DNA-particles. The conjugation and crossing-over, which enter into the structure as the main components, is only the physical manner by means of which occur the changes of the genetic information of the generative cells. Both of these biophysical processes are connected with the wave nature of DNA. Meiosis represents two successive divisions, which end in the formation the four specific cells –gametes. The gametes are characterized by a difference of the DNA-particle, with the same DNA-wave and the haploid set of chromosomes. As compared with the initial diploid cell, they have a different DNA-particle, and the DNA-wave goes down one energy level and the number of chromosomes is reduced by half.

Activating Meiosis

Nature has created another variation of meiosis - activating meiosis. It occurs in many plants – diatoms, some species of green algae, gymnosperms, angiosperms, etc. In higher order plants, this type of meiosis is known as a female meiosis or meiosis during megasporogenesis. In gymnosperms the cytology of megasporogenesis of *Larix sibirica* Lab. was first described in the work of H.O. Juel (1900), which until now has not lost its importance. In higher order plants the succession and spatial organization of meiotic divisions during activating meiosis is identical (Figure 16).

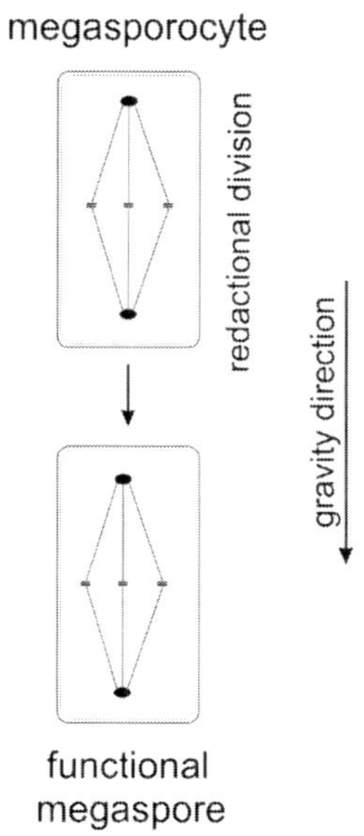

Figure 16. The scheme of a spatial disposition of the karyokinetic spindle during activating meiosis.

After the first meiotic division of megasporocyte, a dyad is formed – a chalazal and a micropylar cell. The spacial position of the spindle from this division coincides with the direction of the polar axis of the megasporocyte and the direction of gravity. Then the micropylar and chalazal cells undergo the second meiotic division. This division of the chalazal (lower and larger) cell has the same polarization and spatial organization as the first meiotic division. The lower generative cell of a tetrad becomes the functional megaspore. It passes to the reproductive process and gives rise to the female gametophyte. The three upper megaspores soon die. If you look closely at the scheme of the spatial organization of this meiosis, it will remind you of another type of cell division – the two divisions of activating mitosis following each other. Therefore, wc can suppose that, megasporogenesis inside of a reproductive complex and generative cells creates such physical conditions, which are simultaneously initiated at the beginning of two processes - of a DNA-article and a DNA-wave change on the basis of the two types of generative cell divisions - a typical meiosis and activating mitosis. The result of this union has been the emergence of a new variation of meiosis – the activating meiosis.

The study of the reasons for transition from a typical meiosis to an activating meiosis will be analyzed on the example of the life cycle of *Pinnularia viridis* (*Bacillariophyceae*). As known all diatoms are unicellular organisms with a diploid cycle of development. On the one hand, the life of pinnularia is organized so that during ontogenesis the cell sizes are progressively decreasing. The individuals (from division to division) are incapable of any change in size due to the rigidity of their siliceous outside shell. The result of the reduction in individuals is the increase of intracellular pressure and viscosity up to a specific limit. As mentioned above, the increased viscosity is the main physical requirement in order for the meiosis in the cell to begin. On the other hand, a pinnularia has the shape of a highly elongated ellipse. As we know, the liquid density increases in a direction of gravity. The spatial organization of the nuclei during the two meiotic divisions coincides with the polar axis of the initial cell and with the direction of gravity. The nuclei have different fates as well. From the four gametes the only viable one is the apical nucleus, the other three are dead. The combination of these two factors (a rising pressure and a polarity meiotic process) becomes a stronger differentiation between the cytoplasmic layers. That is why during a reduction in the size of individuals inside their cells it creates the physical conditions necessary for the transition of the DNA-wave at higher energy levels. Thus, during meiosis are initiated the two different and independent reproductive processes: 1) the change of the DNA-particle of the gametes and 2) the restoration of the energetic potential of one of the gametes or their "rejuvenation" on the basis of a DNA-wave. The principles of the formation of hereditary and energetic programs of the future plant during activating meiosis are shown in Figure 17.

To understand the activating meiosis nature let's assume, that the DNA-wave of the initial mother cell begins the meiotic division on the 3hω energetic level. During radiation the energy of the DNA-wave comes down from the 3hω to the 2hω level. During the first meiotic division, when the homologous chromatids are divergent, the DNA-wave of the apical nucleus moves two levels upwards (from the 2hω to the 4hω). The DNA-wave of a basal nucleus, in accordance with the conservation of energy law, descends to two levels down (from the 2hω to the 0hω). At the same time the DNA-particles of four chromatids by means of conjugation and crossing-over are changed as well. Between the first and second meiotic divisions the DNA-wave radiation does not take place. Therefore, the second meiotic divisions of the apical and basal nuclei begin on, respectively, the 4hω and the 0hω levels. The apical nucleus gives rise to two gametes with a biological energy, respectively, the 8hω and the 0hω levels. The basal cell forms the two gametes, which remain on the 0hω level because the energetic states of a DNA-wave lower than zero level don't exist. The three gametes, which haven't the biological energy, will soon die. The apical gamete, after changes of the DNA-particle and the DNA-wave, begins on the 8hω energy level in the new cycle of ontogenesis.

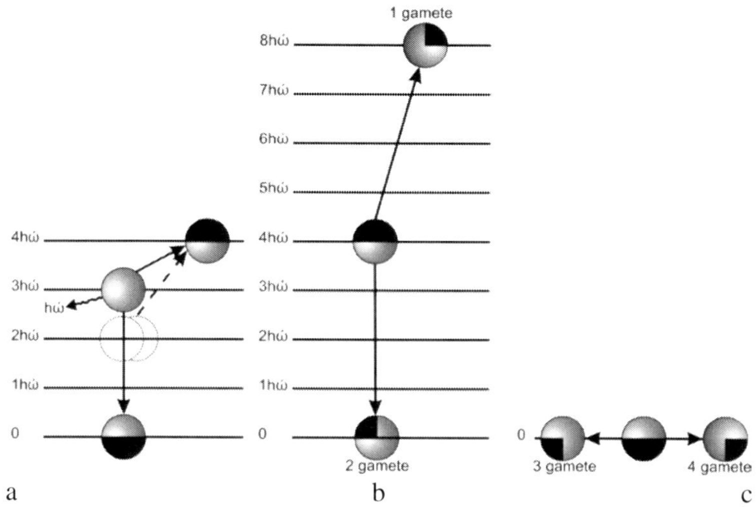

Figure 17. The change the energy during activating meiosis: a) during 1st division; b) during 2nd division of the cell of apical pole; and 3) the cells of basal pole.

Thus, the activating meiosis is a variation of combined mitosis, which unites the two different and independent processes – a typical meiosis and an activating mitosis, leading to a simultaneous change of a DNA-particle and a DNA-wave. This type of cell division is always associated with the sexual process and enters at the reproductive cycles of all species of higher order plants and species of some of the systematic groups of lower order plants. The activation meiosis is characterized by the following characteristics. During the two meiotic divisions the four gametes (or free nuclei) are formed with a haploid set of chromosomes. They are differed between themselves and from the initial mother cell both on a genetic basis (by a DNA-particle) and an energetic basis (by a DNA-wave). These gametes or nuclei have different fates. Only one or two of them (which are "rejuvenated") continue to further reproductive development or pass on to the ontogenesis. The other gametes and nuclei, which during meiosis have lost energy, are dead. The specific physical conditions for the beginning of this meiosis should be established inside of reproductive complex and inside of generative cells. First, a precise polarization in regards to the forces of gravity, the position of initial mother cell and all its derivatives which remain at the reproductive process. Second, a strict spatial orientation as regards to the forces of gravity, the position of the plane division of initial mother cell and its apical derivative. Third, is the increase of intracellular pressure and viscosity. Therefore a cell undergoes the activating meiosis inside the zygote, which is covered by a thick envelope (a Spirogira), or inside a cell with rigid siliceous cell-wall (a Pinnularia), or inside the special reproductive organs, for example, inside the tissue of megasporangium (in higher order plants).

Though many aspects of the nature of mechanisms of cell (or nuclei) divisions, especially in regards to the wave properties, are still unknown, it is obvious that the reproductive cycle of higher order plants include all types of cell divisions, which are connected with the different variants of the change in a DNA-particle and a DNA-wave (Figure 18).

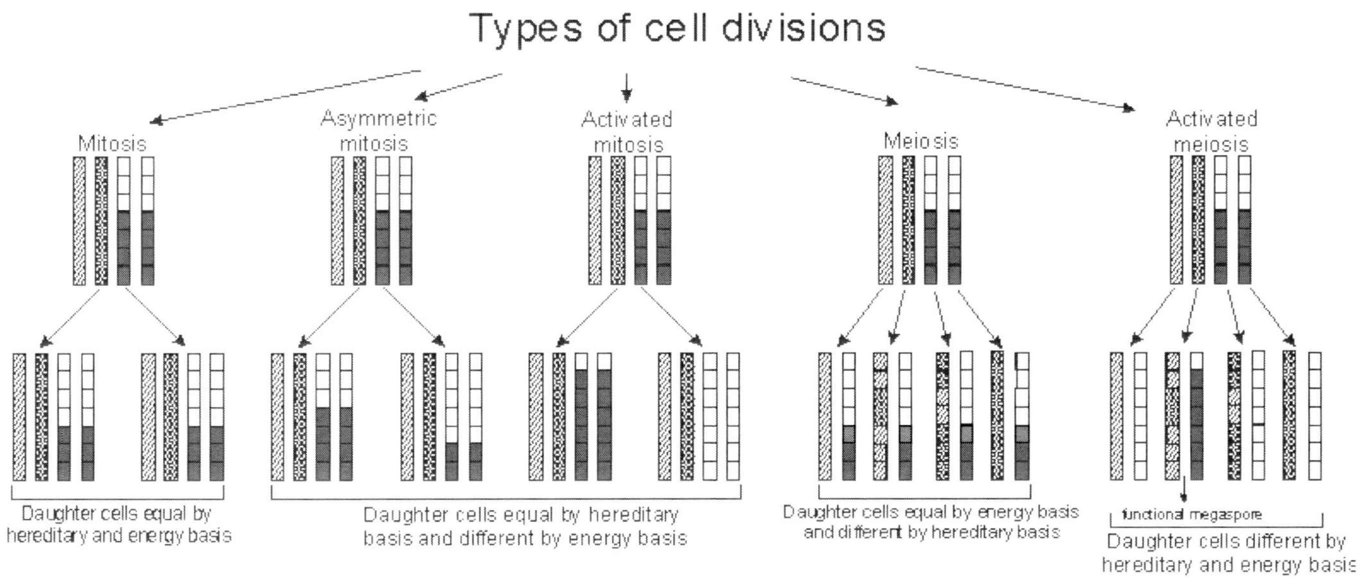

Figure 18. Corpuscular-wave model of the change of DNA-wave and DNA-particle during mitosis, asymmetric mitosis, activated mitosis, meiosis and activated meiosis.

1. Typical mitosis. It gives a beginning to the two equal cells or nuclei (diploid, haploid), which are similar both on a hereditary basis and on an energetic basis. 2. Activating mitosis. It gives a beginning to the two cells or nuclei (diploid, haploid), which are similar on a hereditary basis but are different on an energetic basis. 3. Typical meiosis. It gives beginning to the four haploid cells or nuclei, which are similar on an energetic basis but are different on a hereditary basis. 4. Activating meiosis. It gives a beginning to the four haploid cells or nuclei, which are different both on a hereditary basis and on an energetic basis.

The ability for sexual reproduction is one of the basic functions of living systems, in which the different types of cell divisions play a role. All of them represent a very complex and many-sided physical-chemical process. This information has extended our understanding of the sexual process, proving its corpuscular-wave nature and showing how the different divisions participate in the formation of the dualistic program of future plant development.

Free-Nuclear Mitosis

The next type of cell division in not connected with a change of corpuscular-wave characteristics in a DNA molecule. This free-nuclear mitosis it is the type of mitosis, during which is formed the non-cellular structure which is a multi-nuclear symplast, or a cyncytium, because after the division of the nucleus (mitosis), there is no division of body cells (cytotomy). D. Mazia (1961) identified the two varieties of this type of mitosis: a) a division of nuclei with a delay of cytotomy, and 2) a division of nuclei without the cytotomy. The end result of free-nuclear divisions is the formation of specific multi-nuclear constructions (cells, tissues and organisms), which are characterized by a very large (gigantic) size of cells and nuclei that are in a cytoplasm (parietal layer or continuous mass). The free-nuclear mitosis is widely represented in the plant world: siphonogamous algae, fungi, gymnosperms, etc. It is necessary take note that the symplasts of different species of plants are distinguish by an origin, a type of tissue, a formation stage and reasons for the arising of free-nuclear mitosis. Therefore the every case is requires separate analysis. In this chapter we will study the reasons for the beginning of two varieties of free-nuclear mitosis, namely, when the meiosis has changed to the free-nuclear mitosis.

1) Division of nuclei with the delay of cytotomy. Let us show in an example of the reproductive cycle of *coniferous,* what the conditions are that are necessary for the beginning of free-nuclear mitosis (Singh, 1978; Kozubov et al., 1982; Trenin, 1988). Let us take for this example the two adjacent reproductive processes – a megasporogenesis and a megagametophytogenesis. As known, the female gametophyte originates from a functional megaspore. The functional megaspore takes its origin from the megasporocyte during the two divisions of activating meiosis. The meiotic metaphase chromosomes are smaller than mitotic metaphase chromosomes (Darlington, 1965; Swanson and Webster, 1977; John, 1996). As a result there is more elastic energy contained in them. In addition, the chromosomes are coiling twice – during the 1st and the 2nd meiotic divisions. All this energy, which is liberated during telophase I and II in their uncoiling, enters into the gametes. Because the megaspores contain so much energy they begin to grow before the division. It leads to a change of its nucleocytoplasmic ratio. Soon the three nonfunctional megaspores, with a DNA-wave of zero level, without division are dead. The functional "rejuvenated" megaspore gives rise to the

female gametophyte. The growth of the megaspore forestalls the speed of the biomolecule synthesis (which is a constant). Therefore it begins a vacuolation. The vacuum volume is progressively increased and the cytoplasm occupies in the cell a parietal position. We also note here an interesting fact. Electron microscopic studies have shown that the megasporocyte has two envelopes - an outer and inner primary (Trenin, 1988). The inner layer around the dyad gradually thickens and in addition appears as a callose layer. However the functional megaspore continues to grow, in spite of the resistance of its thickening envelope.

Before the first mitotic division the megaspore is a big cell (Konar and Moitra, 1980; Kozubov et al., 1982; Trenin, 1988). The nucleus is situated at its micropylar end. In the center of cell is placed a large vacuole and the cytoplasm occupies only a thin parietal layer. Therefore during telophase the electromagnetic field, which is created around the two sets of uncoiling chromosomes, covers only the insignificant part from a whole volume of cytoplasm megaspore. The pectic substances in it are not enough to create a formation between the two nuclei of the phragmoplast or a median plate. That is why, after the ending of karyokinesis there will not be a process of cytotomy. The two newly formed nuclei remain in the megaspore and it becomes a 2-nucleus cell. The energy, which during telophase is liberated from the two uncoiling nuclei, enters into a 2-nucleus megaspore. It then causes its further growth. Then follows a strict specific number of free-nucleus divisions (for example their number is 8 in *Taxus*, *Zamia* and *Ginkgo* (Konar and Oberoi, 1969; Kozubov et al., 1982) during which time the free-nuclear female gametophyte is formed.

The growth of the female gametophyte is greatly restrained by sporoderm, nucellus and integument, which are tight to gametophyte and to each other. Therefore during megagametophytogenesis the volume of the nuclei is increased more quickly than the volume of the gametophyte. As a consequence, the nucleocytoplasmic ratio gradually normalizes. At the moment that its normalization begins, the transition from a free-nuclear to a cellular female gametophyte and its development as a cell structure begins.

2) Divisions of nuclei without cytotomy. Another situation is observed in coenocytic alga of *Vaucheria sessilis* with the oogamy sexual process (Komarnitskii, 1975; Life of plant, 1978). In each oogonium which is formed by one ovicell and changed into the antheridium there are many spermatozoids. The oospores, which are forming as the result of fertilization, after a period of rest, begin a new cycle of their own development from the activating meiosis. The meiosis ends by the formation of four nuclei from which survives only one and changes the environment in the oospore. Apparently, the energy, which is liberated during the uncoiling of the meiotic chromosomes, first, assists in the destruction of the thick cell-wall of oospores from within, and, second, evokes the germination of gametes due to the cytoplasmic growth. Therefore the nucleocytoplasmic ratio is changed before the beginning of division of the gamete. For this reason, the first mitotic division of oospore it is a free-nuclear mitosis. The unlimited growth of the thallome threads from outside, leads to the fact, that all free-nuclear mitosis is being ended by a proportional increase both in the nuclear volume and cytoplasmic volume. Thus, at each division of the nuclei the symplast is restored and the nucleocytoplasmic ratio, which was created during the germination of gamete, is changed. And this will occur throughout the entire life of the alga. The body of vausheria gradually develops in a strongly branching thread, and represents a symplast with many free nuclei.

Thus, the free-nuclear mitosis does not change the corpuscular-wave characteristics of the DNA molecule. The main reasons for a beginning of this type of mitosis is that there are morphological peculiarities in the development of meiotic spore and the presence of them after meiosis becomes an abundance of energy, which causes a rapid growth of the meiospore, and changes the nucleocytoplasmic ratio.

Endomitosis

Endomitosis is a type of mitosis where a replication of a DNA-particle takes place in the absence of a cell and nuclear division. The cytology of endomitosis is very different from the scheme of typical mitosis. This variation of mitosis is found only during specific stages of the vegetative and generative organs, tissues, and cells of plant and animal development. The endomitosis during sexual development is revealed in the generative cells, which are the main elements of the reproductive process (for example, the ovicells in conifers) and in the generative cells which are doomed to death (synergids, antipodes, endosperm, haustorium, nutritious cells of oocytes, etc.). This type of mitosis is often connected with specific functional activities of cells (Zhimulev, 1992, 2003). The main signs of endomitosis are: an increase of cell size, a cytotomy absence, a change of nucleocytoplasmic ratio, numerous DNA replication without division or with separation of the chromosomes on the chromatids, a nucleus and nucleolus without terminations, high nucleolar and synthetic activity, and others (Lorz, 1947; Alov et al., 1969; Kiknadze, 1972; Chentsov, 1978; Nagl, 1978, 1982; Zhimulev, 1992, 2003, Weingartner et al., 2004; etc.). For the understanding of the nature of endomitosis let us again turn to the embryology of conifers, beginning from the megagametophytogenesis stage.

The intensity of the growth of a free-nuclear female gametophyte during the first free-nuclear divisions is high (Singh, 1978; Kozubov et al., 1982; Trenin, 1988; Kuznetsova, 2002). The ratio of the volume of megaspore to the volume of cellular gametophyte in a spruce is equal to 1:3500 (Kozubov et al., 1982). During the cellular stage the speed of its growth slows down and then stops. The first cells of the cellular gametophyte are greatly vacuolated. The vacuoles during megagametophytogenesis are now gone, and the cytoplasm is filled by the nutrients (Figure 19-a, b). As we can assume, the reasons for the transition of the gametophyte cells in the endosperm tissue and the initiation of endomitosis may be the following. It is obvious that the numerous divisions of free nuclei, cells and the intensive synthetic processes inside a female gametophyte should lead to an increase of osmotic pressure in it. By the gas laws the same pressure in the confined volume can be achieved by the two ways: 1) by a small number of molecules with a high velocity; and 2) by a large number of molecules with a low velocity. The analysis has shown what the environment should be during a free-nuclear stage of megagametophytogenesis is developing in the first way, and during a cellular stage - in the second way. The speed of biomolecules is directly connected with their ability to enter into a chemical reaction. Thus, the processes of oogenesis and endospermogenesis are proceeding under the conditions of progressively rising intracellular pressure and an increased density of the nucleoplasm and cytoplasm. A direct consequence of the changes in intracellular pressure and speed of biomolecules are: to minimize the metabolic activity of the endosperm cells; cause a biosynthesis of chemically inert substances such as storage proteins, lipids; and a transition of ovicell and endosperm cells to the division by means of endomitosis. The corpuscular-wave nature of endomitosis is inferred from the dualistic scheme of usual mitosis as well. The mechanism of division, apparently, should be the same both in plants and animals.

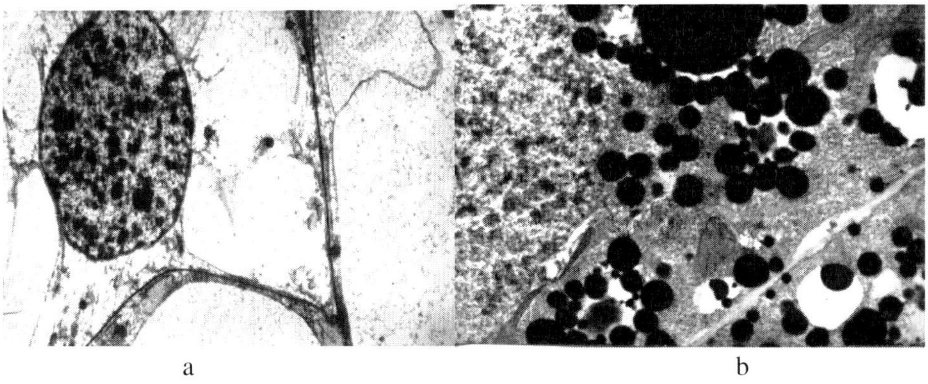

Figure 19. The transformation of ultrastructure of endosperm cells from the beginning (a) to the end (b) of the cellular stage of megagametophytogenesis in *Pinus sylvestris* L.

The DNA-radiation is initiated in a new cycle of the development of two cells due to the formation of two centrioles and a doubling of a DNA-particle. The interval between the radiations of a DNA-wave in the conditions of increasing of pressure becomes less and less until it ceases. The dense nucleoplasm during prophase reduces the rate of the flying away of centrioles, and the electric field between them will expand more slowly. Therefore within its range will be only the central regions of the coiling chromosomes and the distal whose parts begin to unwind. The chromosomes, unwinding, quickly reach the envelope of the nucleus and after that a DNA-wave and a biofield with the two zones of its action – a near one and a far one – are restored. At this moment the centrioles by means of biofield forces are returned into the inner region of the nucleus. Thus the cell, without division, shall enter into the interphase with a double set of chromosomes. The next DNA-wave radiation takes place in the end of the G_2 period, initiating once more the doubling of a DNA-particle and the next endomitosis division of a cell. The endomitosis leads to abundant synthesis of DNA, the formation of polytene chromosomes and the cells with a multiple chromosome complex which have a number of DNA-particles more than 2n. The divisions will continue as long as the DNA-wave radiations are not stopped.

In conifers the processes of oogenesis, embryogenesis and the second part of megagametophytogenesis (endospermogenesis) are entering into a constant volume of the female gametophyte. In them, cells during each division, enters a free energy from two inner sources. Therefore the pressure into the gametophyte continues to grow, but only up to the determinate maximum, namely, until the DNA-wave radiation is not stopped. It is necessary to emphasize, the maximum limits of pressure for the oogenesis, embryogenesis and endospermogenesis are different. Gradually the ovicell, embryo and endosperm (and female gametophyte as a whole) are coming to an equilibrium state at a critical point where they haven't the freedom of degrees. During their maturation the change of the entropy, temperature, pressure and volume are so decreased that they begin to head towards zero. The dissimilation and assimilation processes are minimized as well (Tokin, 1955). Therefore all these reproductive structures cannot by themselves go out from the equilibrium state. Without external influence they are doomed to death.

A high specificity of endomitosis and environment lead to changes in the morphology and metabolism of the generative cells. The energetics of animal eggs during maturation, for example, is provided from three independent sources (two internal and one external). During every division the portion of energy that comes from the 1st and 2nd internal sources turns into a kinetic energy of molecules, which remains in the same cell, and has caused the fast growth of the oocyte. The mature eggs increase in size by hundreds and sometimes thousands of times in comparison with the initial oocyte volume (Raven, 1961; Eisenstadt, 1984). Since growth is mainly carried out due to a cytoplasmic growth of oocyte, accordingly, there is a shift of the nuclear-cytoplasm ratio (Nagl, 1982; Eisenstadt, 1984; Tokin, 1987). For example, in eggs the nuclear-cytoplasmic ratio in a leech *Nephelis* is 0.15/428, and is equal to, 0.3/3 of the somatic cells, i.e., the difference is about 300 times (Tokin, 1987). Perhaps, the metabolism of the oocyte during endomitosis is changing for the same reasons. At every division the chromosomes during their uncoiling produce a large amount of free energy. This energy activates a RNA-polymerase and initiates the processes of transcription and translation in the oocytes. This explains the nature of a number of phenomena, which occur during the maturation of eggs: a high nucleolus activity (Alov et.al., 1969; Ermakov et.al., 1982; Nagl, 1982), the intensity of RNA synthesis (mainly rRNA), enzymes, proteins, etc. (Brachet, 1960; Kiknadze, 1972; Wiegand, 1996; Zhimulev, 2003). However the main part of newly synthesized biomolecules, such as the enzymes, RNA and complex precursors of yolk proteins and lipids have a low velocity and the ability to enter into the chemical reactions and the conditions of a dense nucleoplasm and cytoplasm. That is why the RNA, yolk proteins and lipids being deposited during the processes of transcription and vitellogenesis are the energetic rich and metabolic inert substances. It should be noted that the rRNA molecules are an energetic rich substance as well. The period of the half-life of RNA is increasing, and they become the long-lived molecules (Kiknadze, 1972; Eisenstadt, 1984; Tokin, 1987). As a result, the metabolism of a mature egg is so unchanging that it can be compared to suspended animation (Tokin, 1955). All nutrients are being deposited in the oocyte as fat bodies and RNA and will be use after fertilization by an embryo.

In plants, the phenomenon of polytene chromosomes and endomitosis is less studied than those of animals. In conifers, it occurs during the oogenesis and endospermogenesis processes, as a result of a pressure increase. The ovicell of *Pinus sibirica*, for example, within 2-3 weeks is growing rapidly, it attains a large size and fills up with proteins and lipid drops (Trenin, 1988). During this time the initial cell of archegonium undergoes six cell divisions, from them comes two activating mitoses and four endomitoses. During endomitosis is formed the polytene chromosomes with 32S DNA (Ermakov et al., 1981; Barantseva, 1992; Rees and Sternberg, 2002). For comparison, the rate of division of somatic cells is 24 hours (Nekrasova, 1972). During the process of endosperm formation (from the first functional megaspore division up to the mature endosperm formation) is replaced by three types of cell divisions, namely, free-nucleus mitosis, typical mitosis and endomitosis. As known the female gametophyte of the gymnosperm species is a haploid tissue. The date that the endosperm cells are divided by means of endomitosis are experimentally proved (Lorz, 1947; Fujita, 1961; Pichot and Maâtaoui, 1997). I.e., the endosperm represents the specificity structure with multiple ploidy levels, nuclear fusion, incomplete division, etc. (Nagl, 1978)

The phenomenon of polytene and endopolyploidy in plants is shown in a number of generative cells, such as the antipodes, the embryo suspensor, synergids and tapetum tissue, which during activating mitosis leaves a reproductive process (Nagl, 1976, 1978, 1981; Garvalheira, 2000). All these generative cells are the cells of the basal pole. So they have another fate. They begin to divide by means of endomitosis only if their biological harmonic oscillator has even a small quantity of biological energy. However, after going down up to the 0th level, all these cells and their derivatives will die soon. Often before death these cells are carrying out a specific function, due to the specificity of endomitosis (intensive growth, high synthetic activity, etc.). For example, the cells of the syspensor takes the embryo from the ovicell and moves it into an endosperm, where it is under new conditions (pressure and volume) and will be developed until seed maturation.

Thus, the chief event during endomitosis has changed the number of DNA-particles in the chromosomes. This variation of mitosis leads to the origin of polytene chromosomes, the cells with an abnormal morphology and metabolism. In most cases polygenic cells are not viable (an irreversible endomitosis). The examples of irreversible endomitosis are: a macronucleus in a paramecium, antipodes, the embryo suspensor, synergids, tapetum tissue, endosperm, etc. The cell, which is a survivor of a number of endomitoses, is a fertilized egg of animals and some plant species. It is a reversible endomitosis. In order to survive and take part in fertilization the chromosomes must restore the haploid structure from a polytene one and the ovicell must change its own internal structure. This process as a rule takes place during a special type of cell division, well-known in embryology as a cleavage. We will begin to discuss the corpuscular-wave model of this type of mitosis below.

Cleavage

The formation of a diploid zygote is the end result of fertilization. The beginning of embryogenesis in many organisms begins from the cleavage. Currently, the cleavage is considered only as an early stage of embryogenesis. We have reason to suppose that male and female gametes are distinguished not only by their morphology and metabolism, but also by the state of their DNA-particles and DNA-waves. Therefore let us a look at the given process from another aspect. Namely, that the cleavage is a period of correction of corpuscular and wave differences between the genomes of both gametes. During the study of the cleavage we took into consideration the environment inside the reproductive complex (before and after fertilization), the direction of the change of the DNA-particles and the DNA-wave for the chromosomes of male and female gametes.

It should be noted that the corpuscular-wave model of cleavage begins to coincide with the experimental data after we have assumed that the oscillation frequency of the DNA-wave of a male gamete should be higher than the oscillation frequency of a female gamete. In this case, the chromosomes of mature eggs and spermatozoon (sperm) present very different particles and waves. In such a state their genetic apparatus, even united, is not able to function as a united whole. The new physical conditions at fertilization and the mechanisms of syngamy of a sexual nucleus include the process of changing the DNA-particles and DNA-waves in both gametes. The haploid sets of the chromosomes of the ovicell and sperm after corresponding conversions fuse and give rise to the first true diploid cells or a sporophyte initial, are realized in species-specific plants during ontogeny (Kuznetsova, 2002). Without the fertilization, all male and female gametes as the cells with an abnormal genetic structure must die, that is actually occurring under natural conditions. The ovicells of angiosperms remain viable for 6-7 days (Poddubnaya-Arnoldi, 1976). In Scots pine the irreversible changes of the ovicell ultrastructure appear in 7-10 days after pollination (Kuznetsova, 1994).

What is a fertilization phenomenon of a physics position? Before fertilization the polytene chromosomes of a mature ovicell are a set of closely related DNA-particles. The stability of such a cumbersome structure guarantees the equilibrium state, which will be reached during development of female gametophyte (megagametophytogenesis and oogenesis) as a whole and for its parts (between the ovicells and endosperm tissue). The ovicells are fully formed, when the pollen tube comes up to the female gametophyte. The pollen tubes during growth between a pollen chamber and an ovicell make the "passage", which crosses the tissue of the nucleus, neck cells and sporoderm. The sperm serves as a piercer on the final part of the journey to a nucleus and crosses an ovicell envelope, a layer of cytoplasm (Figure 20-a).

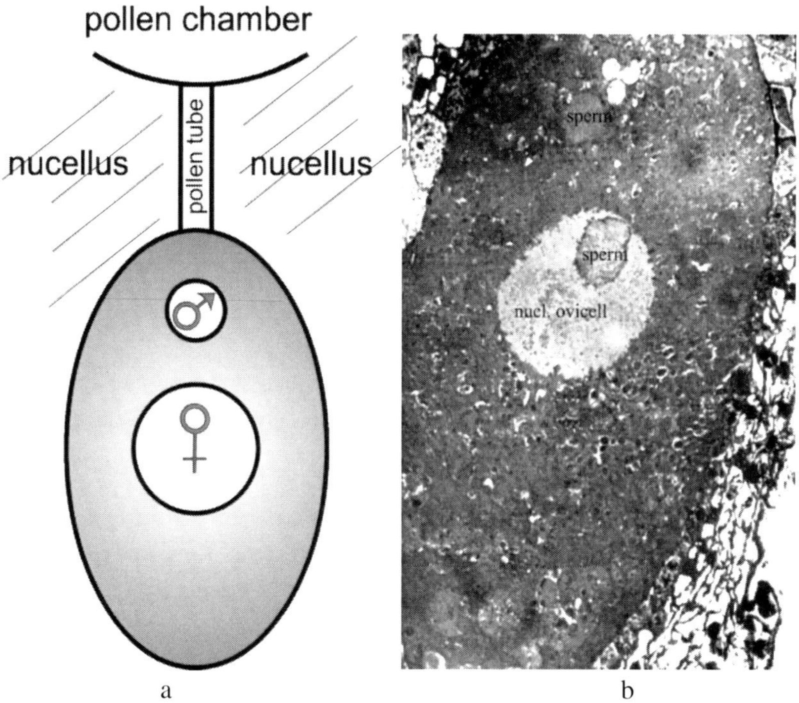

Figure 20. The fertilization in *Pinus sylvestris* L.: the entrance of functional sperm into the cytoplasm (a) and nucleus of ovicell (b) nucl ovicell – nucleus of ovicell sperm – functional and nonfunctional sperms.

At the moment of penetration of the contents of the pollen tube into the cytoplasm of ovicell causes a short-lived contact with the pollen chamber across this "passage". The consequence is, first, an intracellular pressure into a female gametophyte and the ovicell is decreased. Second, the female gametophyte and ovicell during piercing through the sporoderma and ovicell envelope of the pollen tube and sperm have received a blow and lose balance. Third, the sperm by its wave nature is the high-frequency harmonic oscillator. Thus, inside of a united nucleus a syngamy begins acting on the two different harmonic oscillators – the functional sperm and the ovicell nucleus (Figure 20-b).

We can suppose, that the mechanical piercing of an ovicell envelope, is shown in the increase of intracellular pressure and the gamete syngamy and is the main natural factor for leaving the ovicell (or egg) from an equilibrium, causing a destabilization of its genetic apparatus, an activation of biochemical reactions (a stimulation of Ca^{2+}-dependent processes, membrane depolarization, etc. (Carroll and Epel, 1975; Tokmanov et al., 2002)) and an initiation of an orderly disintegration of polytene chromosomes during the cleavage divisions. As described the different forms of parthenogenesis, at which a female gamete (ovicells, eggs) are now developing without fertilization (Poddubnaya-Arnoldi, 1976; Astaurov, 1977; Kaufman, 1983; Tokin, 1987). However, the activation is always preceded by some stimulus, in which the ovicell is placed out of balance. Moreover, the source of this stimulus, apparently, may be different (pressure, temperature, chemical substances, etc.).

The cytological, morphological, biochemical, and other peculiarities of cleavage division are studied mainly in animals and less in plants. This mechanism in all organisms is universal and has a number of general regularities: the shortening of the mitotic cycle, the transition from polytene chromosomes to diploid ones, the normalization of the nuclear-cytoplasmic ratio, etc. Based on the state of the DNA-particles and the DNA-waves in the male and female gametes at the moment of the unification of their chromosomes, the intracellular events may develop in the following way.

Let us define more precisely the details of the sexual process during gamete fusion. It is shown, that the envelope of the sperm nucleus is destroyed at the union of the nuclei (Chesnoy, 1969; Trenin, 1988) or shortly thereafter. According to the experimental data, after the gamete syngamy is visible, two groups of chromosomes enter the zygote nucleus (Singh, 1978; Trenin, 1988). Then, the pairs of chromosomes are attracted to each other, but the true diploid set of chromosomes at this time is not yet formed. Therefore, after fertilization the starting position of the new genome of a zygote is complex. On the one hand, in the nucleus the two groups of chromosomes oscillate, which differs by a DNA-particles and DNA-waves. On the another hand,, the coiling of both sets of chromosomes is carried out by the two centrioles in the ovicell.

This type of mitosis is more correctly called a combined mitosis because the chromosomes of the male and female gametes are involved in it with different programs (Figure 21). As mentioned above, the egg during maturation is passing into a state of rest. The DNA-wave radiation is terminated. The fertilization is given a push, which destabilizes the reproductive complex and the egg goes out of a state of balance. Because the first radiation of the biological harmonic oscillator of the ovicell does not occur immediately, therefore the chromosomes will continue to coil. They have, for realization of this process, all the necessary components – the many-threaded DNA-particle, a DNA-wave and two centrioles, which are formed in the nucleus egg before fertilization. The DNA-wave of the male gametes at this time goes down in energy level and radiates a quantum of free energy. Due to this energy the sperm DNA-particles are doubled and the centrioles are lengthened by a specific number of monomers of actin. Since the frequency oscillation of a sperm DNA-wave is high, the replication duration of a DNA-particle (S-period) is very short. For example, in the first two hours the nuclei of *Drosophila* are divided every 9.6 minutes (Zhimulev, 2003). Then a double set of chromosomes sperm passes the G_2 period (an internal coiling of chromosomes).

After the end of the G_2 period the chromosomes of two gametes simultaneously enter into the prophase of first mitosis.

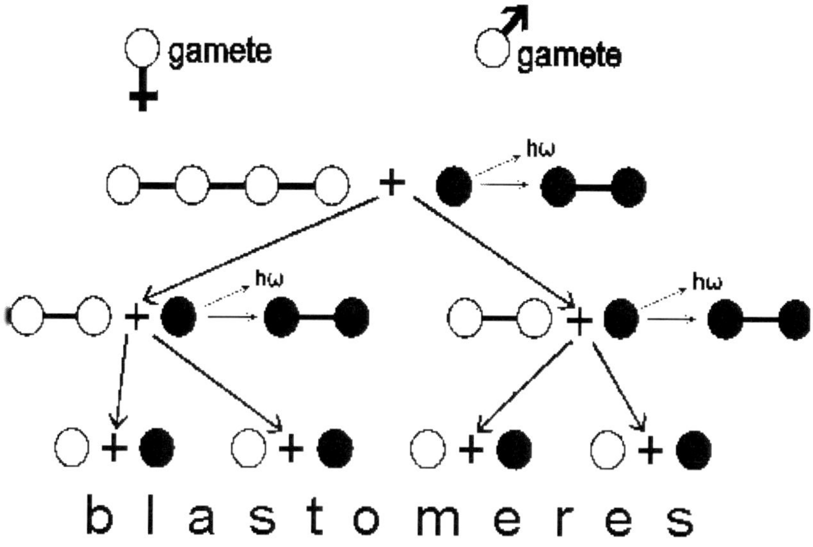

Figure 21. The scheme of the divisions of male and female gametes during cleavage divisions.

The pressure in the fertilized ovicells (the eggs) is reduced just enough for the centrioles to be able to give rise to two mitotic centers, and split in half both a polytene ovicell chromosome (a division without the doubling) and a double sperm chromosome. As a result, the number of DNA-particles in polytene chromosomes after the first division of the zygote is reduced by half.

After the end of the first division of the zygote the two daughter cells (or nuclei) enter into an interphase. Their genome during the G_1 period is represented by the set of polytene chromosomes from a female gamete, whose number has decreased by half and the set of haploid chromosomes from male gamete, the oscillation frequency of which is higher. Therefore, the unstable situation is again originated in the nucleus - the big polytene chromosomes of the ovicell (or egg) oscillate near the haploid high-frequency chromosomes of the sperm (or spermatozoid). The two motions affect each other, causing a transition of the DNA-wave of male chromosomes to a lower energy level. The free energy which is liberated during radiation has initiated the doubling of the DNA-particle, the lengthening of two centrioles and the coiling of both sets of chromosomes during G_2 period. However the process of coiling of high-frequency chromosomes of the sperm occurs faster. The destruction of the sperm chromosome-waves leads to the disintegration of the biological oscillator of zygote, and the set of male and female chromosomes enters into the prophase of the next mitosis. The second cleavage division occurs by the method described above. The chromosomes during cleavage divisions become thinner and shorter (Cleveland, 1949), the nuclear-cytoplasmic ratio normalizes (Raven, 1961; Belousov, 1980, 2005; Tokin, 1987) and the DNA content, which changed insignificantly during the first divisions, begins to increase rapidly (Brachet, 1960). The cleavage divisions will continue, until the chromosomes of the male and female gametes will not emerge on the diploid level of the DNA-particle and on the identical frequency oscillations of the DNA-wave.

The absence of growth and the change of the nuclear-cytoplasmic ratio of embryonic cells are distinctive traits of a cleavage division (Brachet, 1960; Raven, 1961; Belousov, 1980, 2005; Tokin, 1987). The volume of blastomeres up to the end of the period of cleavage even decreases in size (Tokin, 1987). Perhaps, this is connected with the energy state of the embryonic cells. The metaphase chromosomes of cleavage cells are cumbersome and poorly coiled. They are 2-3 times longer than mitotic metaphase chromosomes (Kiknadze, 1972). Therefore the energy, which is liberated during uncoiling in a telophase can cover the energetic costs of only the most vital functions of a cell, such as the metastable state of the cell system, respiration, etc. The decrease in the size of the cells is realized due to the diminishing of cytoplasm volume. This leads to the normalization of the nucleocytoplasmic ratio. The energy deficit has caused an "interruption" of the genetic activity of the divisible generative cells - the lack of nucleolar activity, the synthesis RNA, protein, etc. (Brachet, 1960; Kiknadze, 1972; Tokin, 1987, Belousov, 1980, 2005). Their metabolism and energetics during cleavage divisions are supported by means of a surplus number of the molecules of RNA, enzymes, proteins, lipids, etc., synthesized during the maturation of the eggs (ovicells). All of these biomolecules after a decrease of pressure are activated and incorporated into the intracellular reactions.

Under the totipotency rule is the ability of the cells to give the whole organism its developmental abilities. Be aware that the zygote of animals and plants has the most influence on development. In animals the blastomeres with the regulatory type of cleavage have an equal heredity during the first few divisions of the zygote (Belousov, 1980, 2005; Tokin, 1987). The potency of the cells becomes narrow during embryogenesis and they lose the properties of totipotency. In eggs with a determinate type of cleavage the differences between the blastomeres are shown early as well as the predetermination of its fate as well. The phenomenon of totipotency and differentiation can be explained from the position of the corpuscular-wave organization of genome and corpuscular-wave nature of cell divisions. If the zygote is divided by means of cleavage only, the blastomeres must be totipotent. However, the majority of cleavage divisions occur as a combination of different types of mitoses. It is obvious that the differentiation of blastomeres takes place during the time of the combination of two types of mitosis: a cleavage + an asymmetric mitosis and a cleavage + an activating mitosis. Only the asymmetric mitosis and activating mitosis originate the daughter cells, which are unequal by the wave component of their genome. It should be noted that the process of the differentiation of embryonic cells is strictly a genetically determinate process, in which every species has its own specifics, an order of continuation and a combination of the different types of cell divisions. Therefore the blastomeres in various taxonomic groups of animals differ by the special arrangement inside the eggs and their respective fates.

The zygote of *Pinaceae* undergoes a four cleavage division during proembryogenesis. The first two of them are free-nucleus mitotic divisions and the last two are cell divisions of complex type. The development of the embryo at this time occurs in the ovicell. The nuclei of 4-nuclear proembryos are totipotent and of equal heredity. As known, every one of them can originate viable independent embryos after a cleavage polyembryony. However the third and fourth cell divisions of a zygote are represented by the combination of activating mitosis and cleavage, which leads to a change of the biological energy of the generative cells. Therefore the cells of a 16-cell proembryo lose the property of totipotency and have different fates during embryogenesis. Only cells of the embryonic tier participate in the process of the embryo formation. The rest of the three tiers of cells are dead.

The complex polytene structures of chromosomes during proembryogenesis come back during the diploid state (Ermakov et al., 1981, 1982; Barantseva, 1992). The embryonic cells become smaller. The nuclear-cytoplasmic ratio and the ultrastructure of meristematic tissue are restored (Kuznetsova, 1994, 1998, 2002). The next transition of environment (volume, pressure), which takes place in a proembryo carry from the ovicell into a corrosion chamber, and evokes the uniform response – the simultaneous radiation of chromosome-waves in all four cells of the embryonic tier. Only after that the 4 embryonic cells with a diploid set of chromosomes undergo the first division of ontogenesis, as a whole system.

Thus, the cleavage division is the variation of mitosis, during which the polytene structure of the DNA-particle and the different oscillation frequency of the DNA-wave of the male and female gametes are result in the same type of corpuscular-wave condition of chromosomes. The unification of haploid sets of male and female chromosomes in the diploid sets and the development of the foundation of the multicellular diploid organisms of plants and animals have become possible only after transformation of the genetic material of both gametes during a specific number of cleavage divisions. This variation of mitosis takes place only during sexual development and only in species with a reproductive cycle that includes endomitosis.

Promitosis

Prokaryotes are related to a group of "pre-nucleus" one-celled organisms such as bacterium, cyanobacteria, etc. They are the smallest living organisms with a simple inner structure (Cairns, 1966; Swanson, and Webster, 1977; Shlegel, 1987; Karp, 1996; Alberts et al., 2002; Rees, and Sternberg, 2002; Chentsov, 2004; etc.). The distinctive properties of prokaryotic cells are the absence of the membrane-bound organelles - a nuclei, chloroplasts, mitochondrion, etc. Positioned in the center of the cell is a zone of nucleoid with the DNA molecule. The DNA content in prokaryotes is many times less in comparison to eukaryotes. The sole chromosome represents the circular loop. The DNA is tightly linked to a cell envelope at one point by means of mesosoma, which consists of special membrane proteins. The division of bacterial cells differs significantly from usual mitosis. Its mechanism as a whole resembles the standard scheme of the cell cycle of a eukaryotic cell, but only at the beginning of its evolutionary development. Therefore we have distinguished the manner of the division of a prokaryotic cell as an independent type of mitosis, and have called it promitosis. Promitosis leads to a very exact distribution of genetic material between the daughter cells. The percentage of intrusion is only 0.03% (Chentsov, 2005).

Before proceeding to an analysis of the physical model of promitosis, note a number of traits in which the prokaryotic cells are characterized from corpuscular-wave positions. 1. The cell is an open system. 2. The source of internal energy is only one - a biological harmonic oscillator, or the cell together with the oscillated DNA. 3. The DNA-wave has achieved only the 1st (an excited) energy level, which follows for the 0th (a main) level. Therefore the bacterial cells cannot have a sexual reproduction. 4. The DNA-wave is a standing wave and the energy enters into a cell in two streams. 5. Two synthetic processes are simultaneously initiated during DNA-wave radiation – a replication and transcription, which are not separated by time and intercellular space. 6. The synthetic triple complex during cell cycle is simultaneously functioning in a replication, transcription and translation. The mechanisms of the biosynthesis of DNA, RNA and proteins are similar in prokaryotic and eukaryotic cells. 7. All DNA-particles represents one replicon. 8. The DNA-particle is not coiling during division. 9. The DNA-particle during the cell cycle is firmly connected with the cell membrane by means of mesosoma. 10. The mechanism of cytotomy is not dependent on the time of the termination of DNA synthesis. 11. The duration of DNA-particle replication during exponential phase growth and at the presence of sufficient nutrients is 90-95% of a whole cell cycle. 12. The DNA-wave has direct contact with the cytoplasm due to a lack of a nucleus. Therefore the degree of the reliability and the quality factor (Q) of the biological harmonic oscillator is insufficient.

The macromolecule during the cell cycle changes its state twice: 1) 1 DNA-particle + 1 DNA-wave; and 2) 2 DNA-particles + 1 DNA-wave. The mechanism of promitosis can be presented as the following sequence of events (Figure 22). Each new cell division initiates the DNA-wave radiation. The radiation takes place spontaneously during a period of 20-30 minutes. The DNA-wave lets out a quantum of free energy and moves from the 1st to the 0th (a main) energy level. Established, is that the process of the replication of a circular DNA-particle always starts from the same point as the replication (ORI), which is located in the area of the mesosoma (Jacob et al., 1963). i.e., the place of radiation on the DNA-wave is rigidly fixed. We have assumed, that the beginning of mitosis and promitosis by their nature should be identical. Let's admit further, that the streams of DNA-wave energy beat out the electrons from a mesosoma and this protein becomes positively charged. The excited electrons are passed to the DNA polymerase from a mesosoma, resulting in the enzyme molecules receiving energy and negative charge. At this moment the complex with a charge-transfer will be functioning until the end of the replication process.

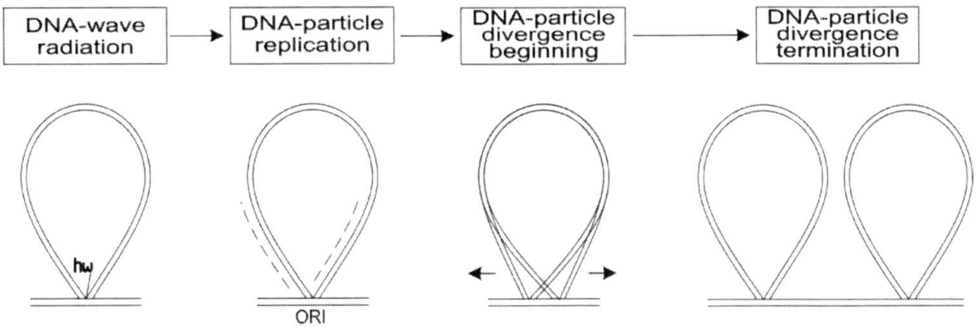

Figure 22. The scheme of the division of prokaryotic cells.

As the DNA-wave is a standing wave, then the energy enters into a cell in a two streams, and simultaneously activates two molecules of DNA-polymerase. The bacterial DNA-particle has a circular structure and a polynucleotide dipolar inner organization in which the negative charges are concentrated in the area of the phosphate groups and positive charges are disposed of in the area of the bases (Wolkenstein, 1983). Molecules of DNA polymerases after receiving the energy and the charge are immediately attracted to the different threads of the DNA-particle and to the different waves of the DNA-wave. The interaction between the molecules occurs in the area of the bases, or where a zone of positive charges of the DNA-particle is situated. Since the enzyme has a negative charge, the electric balance in the place of their unification is destroyed. The two threads of the DNA locally repel and move away from each other, leading to the origination of a DNA-replication fork. The two molecules of a DNA-polymerase from the mesasome (ORI point) begin to passively move along the chromosome on opposite sides together with their own waves. The enzymes during the movement automatically untwist the thread of DNA, carry out the enzyme catalysis and spend their own energy on the reaction of biosynthesis and are reading the information from a contrary wave on the principle of reciprocity. If the DNA-polymerases after a replication end remain in the activation state and the energy is enough for a next round of replication, then they continue a DNA synthesis.

The end of replication is the moment, which initiates the process of cleavage of a double DNA-particle on two daughter macromolecules and the process of cytotomy. The suppression of biosynthesis transpires until the termination of replication breaks the mechanism of cell division. As known, the sole chromosome of all prokaryotes represents one replicon. That is why the paths of two DNA-polymerases during the process of replication must cross in the region of the mesosoma. Because the enzymes are charged negatively, the speed of their motion during approach to each other is slowed, and the speed during divergence, by contrast, is accelerating. At the time of maximum acceleration - a jerk - the mesosoma is torn in half. As a rule, the daughter cells acquire one chromosome. If a division is unequal, then the small cell will be without a chromosome. The mini-cells of E. coli usually reach a size of 0.3-0.5 microns. The newly formed mesosoma, as two positively charged particles, begin to repel and move away from each other. They stop at the distance of ¼ the diameter of the cell (Coulomb repulsion of charges) (Chentsov, 2004). At the same time the formation of a plasmatic membrane occurs in the area between the two mesosoma (Shlegel, 1987). Because the mesosoma are firmly connected with to the DNA-particles, then they split the doubled macromolecule, take away with it copies and the nucleoids completely separate from each other. The main part of the DNA-polymerase up to this time is inactive. The two molecules of inactive enzymes after the collision turn into the cytoplasm, while the free electrons remain on the DNA-particles. The electrons begin flocking to the positively charged mesosoma, their charges are neutralized and the charge-transfer complex ceases to exist.

After the divergence of daughter chromosomes the division of the body of cells begins. The growth of cell at this time slows. Established is that the process of cytotomy is not connected with the mechanism of the DNA doubling (Chentsov, 2004). Most likely the nature of cytotomy is caused by the wave properties of the DNA. As the two biological harmonic oscillators begin to work into a mother cell, around each of them is created their own biofield with two zones of action - the near field (a nucleoid area) and far field (a cytoplasm area). The bilateral motion of the proteins of family *Fts* under the action of biofield forces in the area of the future septum leads to their concentration in a given zone. The cell septum always begins to form from the mesosoma region. Its formation leads to a complete partition between the daughter cells.

Apparently, somewhere after the end of the doubling of chromosomes the next DNA-wave radiation happens in the newly formed cells. The radiation is a purely physical phenomenon, which as such is a spontaneous process and is not connected with the previous cycles of biosynthesis. The liberated free energy initiates the next cycle of replication and a cell division by the method described above. Therefore, if the synthesis of DNA has already occurred in a bacterial cell, then the cycle begins again. As a rule, this occurrence takes place in rapidly growing cultures, where the circular chromosomes with 2-3 intermediate stages of doubling meet frequently (Shlegel, 1987; Chentsov, 2004). The channels, by which the energy from a DNA-wave is utilized by a cell, are the following: 1) a replication; 2) a transcription and translation; 3) a cytotomy; 4) a cell growth; 5) a restitution of ATP from ADP; 6) an excitation of biomolecules; 7) a heat loss and others.

Now let us analyze the algorithm of the action of biological harmonic oscillator and the development of a prokaryotic cell during the cell cycle since the DNA-wave radiation moment. The DNA-wave during radiation falls to a zero (a main) energy level. As a result, the cell energy state increases, and it starts to grow rapidly. All prokaryotes are open systems that are in constant interaction with the environment. Molecules from the environment enter by means of diffusion and active transport (Rees and Sternberg, 2002). The number of molecules increases during biosynthesis as well. On the outside the cell of bacteria and blue-green algae is covered in a rigid cell wall, which is composed from lipids, polysaccharides and proteins (Swanson, C.P. and Webster, 1977; Shlegel, 1987; Rees and Sternberg, 2002). The bacterial wall may be up to 20-30% larger than the dry weight of the bacteria. The rigidity of its framework adds a giant molecule of murein. The rigid cell wall strongly restrains a cell growth and leads to an increase of intracellular pressure according to the gas law (Feynman et al., 1963). The pressure growth causes the transition of the DNA-wave from zero level to the 1hω excited energetic level. When the cell enters into a state of balance with the external environment, its growth ceases. The excited state of the macromolecule is unstable. Therefore it soon returns to the main state (on a 0th energetic level) and the quantum of free energy re-enters into the cell by two streams. This energy of the DNA-wave has initiated the processes of DNA doubling, the restoration of mesosoma sizes, the cells growth and the next division of the cell. The constancy of the duration of the cell cycle in all prokaryotic cells (about 40 minutes) is explained by the identical mechanisms of division and the conditions of its implementation (Shlegel, 1987; Rees, and Sternberg, 2002). I.e., the whole life of prokaryotes proceeds in the equilibrium state of a wave component of genome. The time of their stay on the 1hω energy level is short-lived. Apparently, the return of the harmonic biological oscillator in the balance is the trigger for a new cell division.

The structure, metabolism and behavior of prokaryotic cells are characterized by regularity. Therefore the cells should possess a large enough quantity of energy and have specific channels available for its utilization. In our corpuscular-wave model the two energetic streams function during a cell cycle: centrifugal (in the system of ATP-ADP) and centripetal (based on processing).

According to present notions the energetics of a prokaryotic cell is provided only due to external sources (Swanson and Webster, 1977; Lehninger, 1985; Alberts et al., 2002; Rees and Sternberg, 2002; etc.). Established is that the bacteria and blue-green algae have a variety of ways of extracting and utilizing the energy from external sources. These are "narrow specialists" because they are able to obtain the free energy only by means of external sources (Racker, 1965; Lehninger, 1985). As an exosubstrate the substances differ, including the specific inorganic and organic substances such as amino acids, creatine, etc. Perhaps, at the beginning the external bioenergetics of prokaryotes was based on the anaerobic fermentation, which is catalyzed by a group of enzymes of the glycolysis. Later a part of the species switched from a fermentation to a respiration, as a more effective (an aerobic manner) way to obtain external energy. Blue-green algae and photosynthetic bacteria have mastered photosynthesis, which is based on the catching of light energy and converting it into chemical energy. ATP, or a compound which is used by cells as a source of free energy, is the end product of all the main paths of energy exchange. The action of a centrifugal stream of energy takes a part in providing the internal source of energy - a biological harmonic oscillator. The fact that the prokaryotic cells have an autonomous internal source of energy confirms the occurrence of a long-term cultivation of *E. coli* under the conditions of substrate starvation (Gusev, 2001). Microorganisms during a period of more than two years are able to survive due to the energy of the biological harmonic oscillator and retain the ability to reproduce.

The centrifugal energetic stream, as in eukaryotic cells, operates on the basis of the processing of different types of RNA. The part of the chromosome on which is carried out the biosynthesis of a DNA-particle is small. In bacteria on the rest of the parts of macromolecules are dispersed up to 6-7 identical sequences (≈ 1% of the whole DNA), which are responsible for the synthesis of rRNA. The large molecules of pre-RNA, formed during transcription, soon begin to spontaneously disintegrate into smaller fragments, which correspond to the size of 16S, 23S and 5S (Shlegel, 1987; Alberts et al., 2002; Rees and Sternberg, 2002). The process of RNA splitting is described in the literature as RNA maturation or processing. We noticed that the nature of processing in the main outline resembles such physical phenomenon as an exponential disintegration. The process of exponential disintegration occurs with the liberation of free energy. The probability of this process is always the same, so the centrifugal energetic stream will continue during all cell cycles. The liberated free energy is used mainly on the energetics of the translation process. The number of ribosomes in bacteria is quite large and may be up to 40% of the weight of the cell (Rees, and Sternberg, 2002). Therefore the synthesis of a main set of proteins can occur regardless of the supply sources (Kolter et al., 1993).

Conclusion

According to this concept, the transition from a promitosis to mitosis should occur with the following changes. 1) The appearance of a nucleus and the "closed mitosis" (Swanson and Webster, 1977; Chentsov, 2004). The transference of chromosomes inside of a nucleus essentially reduces heat loss, creates hindrances in the work of an oscillating system and the probability of its destruction from the outside. This has increased the Q-factor and reliability of the biological harmonic oscillator. After that the transition of the DNA-wave to a higher energy level and the use of internal energy on the building of multicellular organisms become possible. 2. The substitution of immovable mesosoma on the mobile analogue – the two centrioles, or the two molecules of polymeric actin. Established is that the actin is absent in prokaryotic cells [51]. In eukaryotic cells the actin enters as a composition of intranuclear microfibrils and makes up one fifth of the volume of all proteins (Actin ..., 2000; Kiseleva and Allen, 2004). 3. The emergence of the 2nd internal source of energy (metaphase chromosomes). Every daughter cell after its appearance begins to receive in telophase an identical DNA-particle, an identical DNA-wave and an equal amount of free energy for their further development. An additional source of internal energy separates the processes of replication, transcription and translation in time and in space. This increases the orderliness of the streams of energy and intracellular metabolism during a cell cycle. 4. The increase of chromosome numbers, their fixation with the envelope of the nucleus at three points. This increases the power and the coordination of the work of the biological harmonic oscillator, accordance with a set of separate chromosome-waves. The volume of genetic information becomes higher in accordance with the size of a DNA-particle. 5. The life of cells has become carried out in two physical systems - an open and a closed. This has increased the exactness of the distribution of genetic material and energy between the two daughter cells. Only on the basis of this scheme of cell cycle and the physical and biological standpoints could originate the even more complicated type of mitosis - the meiosis.

References

Actin: A dynamic framework for multiple plant cell function. (2000). Eds. Staiger C. et al. Dordrecht., Kluwer.

Alberts, B., Johnson, A., Lewis, J., Raff, M., Roberts, K. and Walter, P. (2002). Molecular biology of the cell, 4th edn. Chaffey, N (Ed.), Garland Science, New York.

Alov, I.A., Braude, A.N. and Aspiz, M.E. (1969). *Basics of* cell *functional morphology*. Moscow, Medicina.

Astaurov, B.L. (1977). *Parthenogenesis, androgenesis and polyploidy*. Moscow, Nauka.

Barantseva,L.M. (1992).*Cytophysiological analysis of archegonium development in Pinus sibirica* Du Tour.Moscow, Moscow State University.

Belousov, L.V. (1980). *Introduction to general embryology*. Moscow, Moscow State University.

Belousov, L.V. (2005). *Bases of general embryology*. Moscow, Moscow State University.

Bogdanov, Yu. F. (1991). The homologous conjugation of chromosomes. *Improvements of recent genetics*. pp. 61-134. Moscow, Nauka.

Brachet, J. (1960). The biochemistry of development. Pergamon Press, London, New York, Paris, Los Angeles.

Cairns, J. (1966). The bacterial Chromosome. Sci. Ann., 214, 36-52.

Carroll, E.J. and Epel, D. (1975). Isolation and biological activity the proteases released by sea urchin eggs following fertilization. *Develop. Biol.*, 44, 22-32.

Chentsov, Yu.S. (1978). *General cytology*. Moscow, Moskovvski universitet.

Chentsov, Yu.S. (2004). *Introduction to cell biology*. Akademkniga, Moscow.

Chesnoy, L. (1969). Sur la participation du gamete male a la constitution du cylasme de l'embryon chez le Biota orientalis Endl. Rev. cytol. et boil. veg. 32(3/4), 273-294.

Cleveland, L.R. (1949). Hormon-induced sexual cycles on flagellates. I. Gametogenesis, fertilization and meiosis in Trichonympha. *Morphol.*, 85, 197-296.

Clowes, F.A.L. (1971). Cell organelles and the differentiation of somatic plant cells: Origin and continuity of cell organelles. *Results and problems of cell differentiation*. (Springer-Verlag, pp. 43-72). Berlin, Heidelberg, New York.

Cytology and genetic of meiosis. (1975). Eds. V.V. Hvostova and Yu.F. Bogdanov, Moscow, Nauka.

Danilova, M.F. and Kashina, T.K. (1980). Apical meristem. *Atlas of the ultra structure of plant tissues*. (Ed. M.F. Danilova and G.M. Kozubov, pp. 17-40). Petrozavodsk, Nauka.

Darlington, C.D. (1931). *Meiosis. Biol. Res.*, 6, 212-264.

Darlington, C.D. (1965). *Cytology*. J. and A. Churchill LDT, London.
Drubin, D.G. and Nelson, W.J. (1996). Origins of cell polarity. *Cell*, 84, 335-344.
Dure, A.F. (1976). The visible events of mitotic cell division. *Cell division in higher plant*. pp. 49-110.Yaoman M. ed., New York, Academic Press.
Eisenstadt, T.B. (1984). Cytology of oogenesis. Moscow,Nauka.
Embryology of Angiosperms. (1984). Ed. B.V. Johri. Springer Verlag. Berlin, Heidelborg, New York, Tokye.
Ermakov, I.P., Barantseva, L.M. and Matveeva, N.P. (1981). Cytochemical study of DNA in the process of oocyte maturation and early embryogenesis of *Pinus sibirica* Du Tour. *Rus. J. Ontogenesis, 12(4)*, 339-345.
Ermakov, I.P., Barantseva, L.M. and Matveeva, N.P. (1982). The structure of the nucleus of developing archegonium and proembryo of Pinus sibirica Du Tour. *Rus. J. Ontogenesis*, 13(6), 611-620.
Feynman, R.P., Leighton, R.B. and Sands, M. (1963). The Feynman Lectures of Physics. vol. 1-3. Wesley, A. (Ed.), Readigg, Massachusetts, Palo, Alto, London.
Fowler, J.E. and Quatrano, R.S. (1995). Cell polarity, asymmetric division and cell fate determination in brown algae zygotes. *Seminars Dev. Biol.*, 6, 347-358.
Fujita, T. (1961). Multinucleate endosperm cells in Cephalotaxus drupacea Siebold ed Zuccarini. *J. Jap. Bot.*, 36, 29-32.
Galitsky, V.A. (2005). Emergence of eukaryotic cells and apoptosis origin. Tsitology, 47(2), 103-120.
Gallagher, K. and Smith, L.G. (1996). Asymmetric cell division and cell fate in plants. Curr. Opin. Cell. Biol., 9, 842-848.
Gamaley, Yu. V. (1972). *Cytological basis of differentiating xylem*. Moskow, Nauka.
Garvalheira, G.M.G. (2000). Plant polytene chromosomes. *Genet. Molec. Biol.*, 23(4), 1043-1050.
Gavrilova, O.V. and Rudakova, E.E. (2003). Functional inequality of nuclei in the thallome of the apical growing algae of *Vaucheria sessilis. Tsitology,* 45(120), 1197-1201.
Gilbert, S.F. (2010). Developmental biology, 9th ed., Sinauer Associates, Inc., Sunderland, Massachusetts.
Greenberg, J.T. (1996). Programmed cell death: a way on life for plants. Proc. Natl. Sci. USA, 93, 12094-12097.
Gurvich, A.G. (1977). *Selected works*. Moscow, Meditsina.
Gusev, V.A. (2001). About the energysource for preservethe viability andamplificationof heterotrophicorganismsin the conditionsof substrate hunger.I. Formulation ofhypothesis. Russian J. Biophysics, 46(3), 862-868.
Heilbrunn, L.V and Wilson, W.L. (1948). Protoplasmic viscosity during mitosis in the egg of Chaetopterus. *Biol. Bull.,* 95(1), 57-68.
Heilbrunn, L.V. (1956). The dynamics of living protoplasm. New York, Academic Press.
Hughes-Schrader, S. (1943). Polarization, kinetochore movements and bivalent structure in meiosis of male mantids. Biol. Bull, 85, 265-300.
Ivanovskaya E.V. (1983). Cytoembryological research of a cell differentiation of plants. Moscow, Izdatelstvo MGU.
Jacob, F., Bronner, S. and Cuzine, F. (1963). On the regulation of DNA replication in bacteria. Cold Spring Harbor Symp. *Quant Biol.,* 28, 329-348.

John, B. and Lewis, K.R. (1965). The meiotic system. Protoplasmatologia VI. (Springer Verlag. pp. 98-183). Wien, New York.

John, B. (1990). Meiosis. Developmental and Cell Biology, Ser. 22, Melburn, Cambridge: Cambridge Univ. Prcss.

Juel, H.O. (1900). Beiträge zur Kenntniss der Tetradentherlung. *Jb. Wiss. Bot.*, 35(15-16), 626-638.

Karp, G. (1996). *Cell and molecular biology: concepts and experiments.* John Willey and Sons Inc. New York.

Kaufman, M.H. (1983). *Early mammalian development: parthenogenetic studies.* Cambrige, Cambrige University Press.

Kiknadze, I.I. (1972). Functional organization of chromosomes. Leningrad, Nauka.

Kiseleva, E. and Allen T.D. (2004). Nuclear pores and nuclear matrix: structure and interaction. Genetics in XXI century: current state and prospects of development. Mat. III Congr. VOGiS (Moskow, 6-12 Aug. 2004). V. 2, Moscow, 215.

Knoblich, J.A. (2001). Asymmetric cell division during animal development. *Nature Rev. Mol. Cell Biol.*, 2, 11-20.

Kolter, R., Siegell, D.A. and Torno, A. (1993). The stationary phase of the bacterial life cycle. Ann. Rev. Microbiol., 47, 855-874.

Komarnitskii,N.A.,Kudryashov, L.V. and Uranov, A.A. (1975).*Systematics of plants.* 7th ed. Mascow, Prosveshenie.

Konar, R.N. and Moitra, A. (1980). Ultrastructure, cyto- and histochemistry of female gametophyte of gymnosperms. *Gamete Res.*, 3(1), 67-96.

Konar R.N., and Oberoi, Y. (1969). Recent work on reproductive structures of living conifers and taxads – a revive. *Bot. Res.*, 35(2), 89-117.

Kozubov, G.M., Trenin, V.V., Tihova, M.A. and Kondratieva, V.P.(1982). Reproductive structures of gymnospermae (comparative description). Leningrad, Nauka.

Kushev, V.V. (1971). M*echanisms of genetic recombination.* Leningrad, Nauka.

Kuznetsova, N.F. (1994). Electron-microscopic study of fertilization and proembryogenesis in Scots pine. *Tsitology,* 36(9/10), 986-992.

Kuznetsova, N.F. (1998). Activation in the course of reproductive development in female generative sphere of Pinus sylvesrtis L. *Tsytology.* 40(10), 849-854.

Kuznetsova, N.F. (2002). The reproductive development of initial sporophyte of Pinus sylvesrtis L. *Tsitology*, 44(2), 159-165.

Kuznetsova, N.F. (2003). The formation of the sporophyte initial of Pinus sylvestris L. as deduced from the deciphering of the reproduction cycle of the pine. Mat. XI s'ezd RBO, vol. 2. (Barnaul, 18-22 august 2003), 149-150.

Lawler, M.L. and Brun, Y.V. (2006). A molecular beacon defines bacterial cell asymmetry. *Cell,* 124, 891-893.

Lee, R.E. (2008). *Phycology.* 4th ed. Cambridge, Cambrige University Press.

Lehninger, A.L. (1985). The principles of biochemistry. V. 1-3. Moscow, Mir.

Life of plants: in 6 vol. (1978).I.V. Grushevitskii and S.G. Zilin eds. Moscow, Prosveshenie.

Lin, H. and Schagat, T. (1997). Neuroblasts: a model for the asymmetric division of stem cells. *Trends in Genet.,* 13, 33-39.

Lorz, A.P. (1947). Supernumerary chromonemal reproductions: Polytene chromosomes, endomitosis, multiple chromosome complex, polysomaty. *Bot. Rev.*, 13(10), 597-624.

Mann, D.G. (1993). Patterns of sexual reproductions in diatoms. *Hydrobiologia,* 269-270(1), 11-20.
Mazia, D. (1961). Mitosis and the physiology of cell division. Ed. J. Brachet and A. Mirsky V. III. New York and London: Academic Press.
Merkis, A.I. (1990). *The force of gravity in the processes of plant growth,* Moscow, Nauka.
Monachova, M.A. (1990). Cytogenetic aspects of space organization of interphase nucleus. *Uspekhi sovremennoi biologii, 1*0(5), 163-179.
Monachova, M.A. (2004). System organization of geno type in the pheno type of cell nucleus. Genetics in XXI century: current state and prospects of development. Mat. III Congr. VOGiS (Moskow, 6-12 Aug. 2004). V. 2, Moscow, 293.
Nagl, W. (1976). The polytenic antipodal cells Scilla bifolia: DNA replication pattern and possibility of nucleolar DNA amplification. *Cytologia*, 14, 165-170.
Nagl, W. (1978). *Endopolyploidy and polytene in differentiation and evolution*. Horth-Holland Publishing, Amsterdam, New York, Oxford.
Nagl, W. (1981). Polytene chromosomes of plants. *Int. Rev. Cytol.*, 73, 21-53.
Nagl, W. (1982). Cell growth and nuclear DNA increase by endorepro duplication and differential replication. Cell growth. (Ed. C. Nicolini, pp. 619-651). New York, Plenum Press.
Nekrasova, T.P. (1972). *Biological bases of the seed-production of* Pinus sibirica. Novosibirsk, SO AN SSSR.
Nelson, W.J. (1992). Regulation of cell surface polarity from bacteria to mammals. *Science,* 258, 948-955.
O'Brien, I.E., Murray, B.G., Baguley, B.C., et al. (1998). Major changes in chromatin condensation suggest the presence of an apoptotic pathway in plant cells. *Cell Res.*, 241, 41-54.
Page, S.L. and Hawley, S. (2004). The genetics and molecular biology of the synaptonemal complex. *Annu. Rev. Cell Del. Biol.*, 20, 525-558.
Petrov, D.F. (1973). *The cytological basis of heredity*. Moscow, Kolos.
Pichot C.E. and Maâtaoui, M. (1997). Flow cytometric evidence for multiple ploidy level in the endosperm of some gymnosperm species. *Theor. Appl. Genet.*, 94, 865-870.
Poddubnaya-Arnoldi, V.A. (1976). *Cytoembryology of angiosperms plant. Principles and perspectives.* Moscow, Nauka.
Racker, E. (1965). *Mechanisms in bioenergetics*. Acad. Press, New York, London.
Raven, Chr. P. (1961). Oogenesis. The storage of developmental information. Pergamon Press, New York, Oxford, London, Paris.
Rees, A.R. and Sternberg M.J.E. (2002). Introduction to molecular biology: From cells to atoms, Moscow, Mir.
Reyvn, P., Ewert, R. and Aykhorn, C. (1990). *Modern botany*. V. 1, Moscow, Mir.
Rhoades, M.M. (1961). *Meiosis. The Cell*. V. III. (Eds. J. Brachet and A.E. Mirsky. pp. 1-76). New York: Academ. Press.
Roegiers, F. and Jan, Y.N. (2004). Asymmetric cell division. *Curr. Opin. Cell. Biol.*, 16, 195-205.
Roth, T.F. and Ito, M. (1967). DNA dependent formation of synaptonemal complex at meiotic prothase. Cell Biol., 35, 245-264.
Shlegel, G. (1987). *General Microbiology*. Moscow, Mir.

Singh, H. (1978). Embryology of Gymnosperms. Encyclopedia of plant anatomy, X (2). Berlin, Stuttgart: Gebruder Borntraeger.

Sounson, K., Merts, T. and Yang, U. (1969). *Cytogenetic*. Moscow, Mir

Stephens, C., Jenal., U. and Shapiro, L. (1995). Expression of cell polarity during Caulobater differentiation. *Seminars Dev. Biol.*, 6, 3-11.

Sterling, C. (1963). Structure of a male gametophyte in gymnosprms. *Biol. Rev.*, 38, 167-203.

Swanson, C.P. and Webster, P.L. (1977). The cell, 4th edn. Prentice-Hall Inc., Englewood, Cliffs, Jersey.

Swanson, C.P. and Young, W.J. (1965). *Chromosome reproduction in Mitosis and Meiosis. Reproduction: Molecular, Subcellular and Cellular*. (Ed.M. Locke, pp 107-129). New York, Acad. Press.

Sybenga, J. (1975). *Meiotic configurations*. Springer-Verlag, Berlin, Heidelberg, New York.

Singh, H. (1978). *Embryology of Gymnosperms*. Encyclopedia of plant anatomy X (2). Berlin, Stuttgart, Gebruder Borntraeger.

Tairbecov, M.G., Klimovitskiy, V.Ya. and Oganov, V.S. (1997).The role of gravity in the evolutionof living systems (biomechanical and energeticaspects). *Biology Bulletin, 5*,517-530.

Terskikh, V.V., Vasiliev, A.V. and Vorotelyak , E.A. (2007). Polarization and asymmetric division of atem cells. Tsitology, 49(11), 933-938.

Terskikh, V.V., Vasiliev, A.V. and Vorotelyak , E.A. (2009). Asymmetriccell division andneoplastic growth. *Biology Bulletin, 4,* 389-394.

Tokin, B.P. (1955). *Immunity of embryous*. Leningrad, Nauka.

Tokin, B.P. (1987). *General embryology*. 4th ed., Moscow, Visshaya shkola.

Tokmanov, A.A, Sato, K.I and Funami Y. (2002). Signal cascade during fertilization. *Tsitology,* 44(3), 227-234.

Trenin, V.V. (1986). *Cytoembryology of Larix*. Leningrad, Nauka.

Trenin, V.V. (1988). *Introduction in coniferous cytoembryology*. Petrozavodsk, Karelsky filial AN SSSR.

Volkenshtein, M.V. (1983). *Theoretical biophysics*. Nauka, Moscow.

Weingartner, M., Criqui, M., Meszaros, T., et al. (2004). Expression of a nondegradable cyclin B_1 affects plant development and leads to endomitosis by inhibitiny the formation of a phragmoplast. *The plant Cell*, 16, 643-657.

Westengard, M. and Wettestein, D. (1966). Studies on the mechanism of crossing-over. III. On the ultrastructure of the chromosomes in *Neottiella rutilans* (Er.) Dennis. *Compt. Rend. Trav. Lab. Galsberg, 35(10),* 261-286.

Westengard, M. and Wettestein, D. (1970). Studies on the mechanism of crossing-over. IV. The molecular organization of the synaptonemal complex in *Neottiella (*Cooke*)saccardo (Ascomycetes).Compt. Rend. Trav. Lab. Galsberg, 37(11),* 239-268.

Westengard, M. and Wettestein, D. (1972). The synaptonemal complex. Ann. Rev. Genet., 6, 71-110.

Wiegand, M.D. (1996). Composition, accumulation and utilization of yolk lipids in teleost fish. *Rev. Fish. Biol. Fish.*, 6, 259-286.

Zavarzin, A.A., Harazova, A.D. and Molitvin, M.N., (1992). *Cell biology: General cytology*. St. Petersburg, Izdatelstvo SpbGU.

Zhimulev, I.F. (1992). Polytene chromosomes: morphology and structure. Novosibirsk, Nauka.

Zhimulev, I.F. (2003). *General and molecular genetics*. Novosibirsk, Sibirskoe universitetskoe izdatelstvo.

Zosimovskaya, A.I. (1973). Mechanisms controlling the entry of cells into mitotic cycle as studies using specific inchibiters of macromolecular synthesis. *The cell cycle. Problems of regulation*. (Ed. O.I. Epifanova, pp. 104-133). Moscow, Nauka.

Chapter IV

Prokaryotic and Eukaryotic Cells: A Vital State and Principle of Development

Abstract

A cell of all prokaryotes, in order to a live, should always be in a process of division. The zero energy level is the primary level for its biological harmonic oscillator. The stability of an oscillating system in equilibrium state is that of a physical body, and the reliability of it, as of a biological body, provides the periodic transition of the DNA-wave (once in an every cell cycle) from the 0th to the 1st energy level. A returning to the initial level is the starting mechanism for the doubling of DNA and for the next cell division. The number of vital states, in which a prokaryotic cell can meet in nature, is three – dividing, resting and dead. The basic vital state is a dividing cell.

The numbers of vital states which eukaryotic cells can have during its individual life are four or five depending on the species of plant. The rise of additional vital states leads to an increase in the duration of the life time of a cell line on the basis of the DNA-wave (from a zygote or its derivatives and up to the natural death of the cells). 1. Proliferous cell. The energetics of a proliferous cell is provided by 3 or 4 independent sources: in plants – 2 internal (1st – the energy of the DNA-wave and 2nd – the energy of metaphase chromosomes) and 2 external (the energy of photosynthesis and respiration); in animals – the same 2 internal and 1 external (the energy of respiration). The processes of replication and transcription are divided in time and in space and use the energy from different internal sources: a replication from the 1st source; and a transcription from the 2nd source. DNA-wave radiation initiates a DNA-particle doubling and a cell division. 2. Differentiated cell. A transition of a cell from proliferation to irreversible differentiation is caused by DNA-wave damping during cell divisions. The process of the state of change is long enough. During this period disappearing from a cell cycle is the reaction of the biosynthesis of the DNA, which is necessary for obtaining free energy. A cell loses the ability to divide and, accordingly, loses the 2nd internal source of energy. The energy from the 1st inner source is used on an energetic securing of the transcription process.

Life without a cell is absent. The statement, that a cell is an elementary and indivisible unit of living matter is the fundamental conclusion following from cell theory. Its development submits to specific regularities which have been formulated by Mazia (1961).

First, growth of a separate cell is limited in space. Second, the life duration of a separate cell is limited in time (this applies to "long-living spores" as well). Both these restrictions are overcome by means of cell reproduction which is a mandatory attribute for life.

The cell during evolution has gone through a long development. Many stages of this evolutionary development are not known. The general tendency of the changes, which were taking place at transition from the prokaryotic to eukaryotic cell, is only hinted at. Namely that during its evolution, the DNA-particle, the DNA-wave, the sources of internal and external energy, the channels of intracellular energy utilization, nucleus, cytoplasm and organelle ultrastructure, etc., have become more complicated and perfected. To proceed from the quantum-mechanical nature of cell, the most important change is the appearance of a nucleus and the transition of a DNA-wave into its volume.

The purpose of this chapter is the description of the vital states in which a cell of prokaryotes and eukaryotes meet in nature, to try and establish the reasons behind a cell system transition from one state to another and to find the regularities (a principle of development of a cell) or general plans, which are inherent in each condition of the cell system. The materials in this sphere of knowledge are very extensive (Mazia, 1961; Swanson, and Webster, 1977; Freifelder, 1983; Shlegel, 1987; Davis et al., 1990; Wolfe, 1993; Karp, 1996; Alberts et al., 2002; Rees, and Sternberg, 2002; Chentsov, 2004; Allen, 2008; Gilbert, 2010; etc.). In our research we put the emphasis on the wave nature of the phenomena.

The above-mentioned regularities are clear not only for the unicellular, but for the multicellular as well. As is known, the development of all multicellular organisms begins from one cell – a zygote, which is formed as a result of syngamy, a male and female gamete. The ontogenesis is a corpuscular-wave process during which a DNA-particle and a DNA-wave are changed independently of each other. The chemical formula of a DNA-particle does not change. This important conclusion is experimentally confirmed many times. The sequence of the DNA nucleotides even after cell death is not broken. Recombination of genes takes place only during reproductive development by means of a special type of cell division – the meiosis and the special mechanism – the crossing-over. The crossing-over, as it has been shown in chapter 3, is a very complex corpuscular-wave process. On the contrary, the energy condition of the DNA-wave during ontogenesis, according to the second law of thermodynamics, changes from a maximum quantity of biological energy (a zygote energy level) to zero (the energy level of natural death of the organisms). I.e., a stock of biological energy during each division decreases one level. Perhaps, the nature of a cell differentiation, their ageing and death is caused by the wave properties of a DNA molecule. Eukaryotic cells of one line during differentiation pass in succession the four different life states – proliferous, differentiated, equilibrium and dead. In the beginning they lose the property of totipotency, after that an ability to divide, and in the end the ability to support life.

The different states of cells can be characterized as a system in which various events and processes are ordered in space and time. A comparison of the scheme of the cell cycle of a prokaryotic cell with the scheme of a proliferous eukaryotic cell shows that they have different principles of development in space. We have also found that the principles of development of a eukaryotic cell depend on a position of the tissue in the body of the plant. For example, the principle of development of the meristem cell is "a wave", and a dead cell of the cortex is "a point". All of this gives us the opportunity to see the same biological facts in a new light.

Prokaryotic Cell

As it has been shown in Chapter 3, the cells of all prokaryotic organisms are open systems and constantly changing in an environment of energy and substance. A biological harmonic oscillator has a rather simple construction and a low energy capacity, namely the oscillator is a cell, the largest part of the cell cycle proceeds on a base (zero or equilibrium) energy level and the state of 0th level is reversible. A cell has a more simple intracellular structure without nucleus, chloroplasts, mitochondrion, or Golgi apparatus. The cell multiplies by means of only one type of cell division – a promitosis. Sexual reproduction is impossible it proceeds from a quantum nature of this process. From the outside a cell is surrounded with a cell wall (pectic envelope, capsule, slime), which fixes its form, and provides protection and rigidity and the double bilayer of a plasmatic membrane, which is characterized by a selective permeability. The presence of a permeable membrane, the rigid cell wall and the quantum mechanical nature of the cells give an understanding of how the biological harmonic oscillator supports its own oscillations on the zero energy level during the cell cycle and in the time of its replacement. At the expense of what energy is replenished, a heat loss occurs in the DNA-wave and it becomes the undamped wave.

Prokaryotic cells in nature meet in the three vital states – basic (dividing cell), at rest (endospores, etc.), and dead.

The Basic State - Dividing Cell

As known, all prokaryotes intensively multiply in a favorable environment. The same intracellular event sequence is carried out during every cell division, which hasn't changed over many millennia. The essence of them consists of the following (Figure 23). The genetic apparatus of a cell works in two regimes: 1) 1 DNA-particle – 1 DNA-wave; 2) 2 DNA-particle – 1 DNA-wave after DNA doubling. The main energy level on which proceeds the largest part ($\approx$ 90%) of a cell cycle is 0th. Only one internal source of energy (the energy of biological harmonious oscillator) and the different external sources (on the basis of fermentation, respiration and photosynthesis) participate in the microorganism energetics, which can obtain free energy from various organic substances (hemotrophs) and sunlight (phototrophs). A cell together with an oscillating DNA-wave serves as the biological harmonious oscillator. It is a surprisingly precise mechanism from a standpoint of physics. Therefore it appears that we are dealing with a biological kind of "perpetuum mobile". But this is not true. The principle of the oscillator action and its "additional charge" in many respects depends upon two factors: 1) the osmotic pressure which varies, as we assume, the strict determinate path during a cell cycle; 2) the ability of the DNA-wave to pass from the zero on the 1st energy level in a condition of higher pressure. All cells carry out these processes of osmoregulation. The violation of it leads to a cell death. Maintaining the oscillations of a macromolecule and stability of zero level are secured by means of the free energy, which enters into a cell from internal and external sources. Otherwise the oscillations of DNA-wave will fade very quickly under the action of friction forces. The duration of the cell cycle and all intracellular processes are closely associated with the work regime of a biological harmonic oscillator. The periodically radiation of the DNA-wave initiates the processes of replication, transcription and cell doubling. Therefore the vital state of a "dividing cell" is the main condition for preservation of DNA-wave function and the viability of a prokaryotic cell. To confirm this statement again turn to the life cycle of prokaryotes on the example of Escerichia coli bacteria, which are the most studied of all the prokaryotic microorganisms.

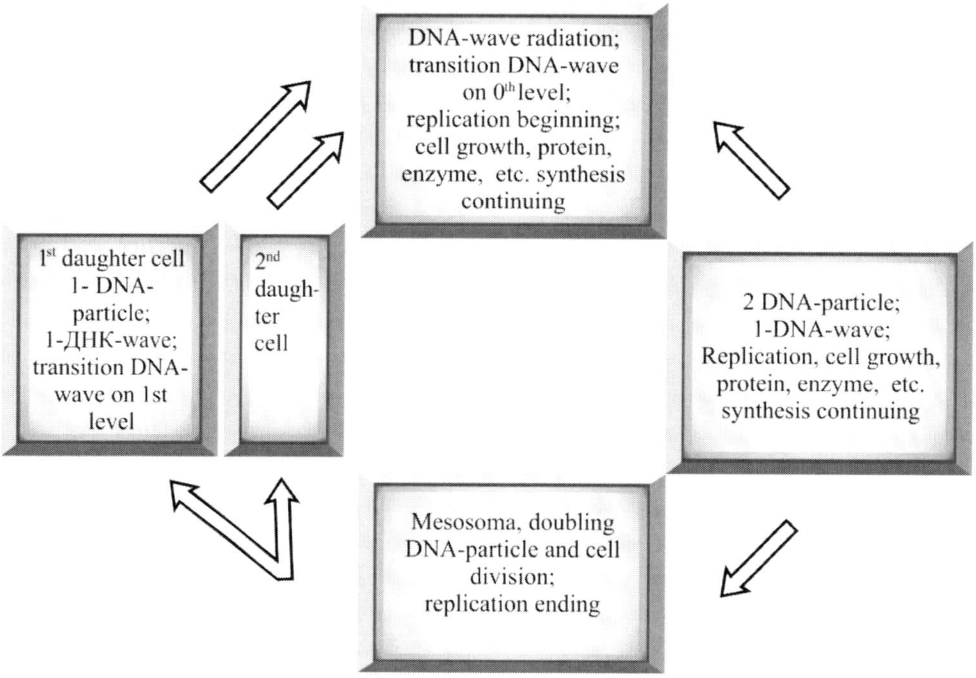

Figure 23. The scheme of the development of prokaryotic cells.

E. coli is a rode-like gram-negative bacterium, which is characterized by mobility and absent of spores. Let us assume, that its chromosome is a particle and a wave and a cell that is the quantum-mechanical system at the same time. In this case the standard scheme of the life cycle of this bacteria and experimental data will have another interpretation. The formation of cells in every new generation begins with the DNA-wave radiation, the return of biologic harmonic oscillator on the 0th energy level and the entering of two streams of free energy into a cell. This energy simultaneously initiates two processes – replication and transcription, in which a DNA-particle and a DNA-wave are involved. The cell is an open system. It has been in contact with the environment during the whole cell cycle. A large number of different inorganic ions, organic and inorganic molecules have entered and been removed from a cell with the participation of the mechanisms of active (including biofield forces) and passive transport of a substance through a membrane. Occurring at the same time, is the intensive biosynthesis of proteins, enzymes, lipids and other biomolecules, which remain in the cell because the plasmatic membrane is impermeable for them. Therefore, the concentration of them in a cell volume during the cell cycle will progressively increase. As known, the concentration of molecules is a function of osmotic pressure, which brings an increase of pressure on the outer envelope. A cell continues to grow up to its determinate size, in so far as the stretching of rigid cell wall, which consists of several layers, is possible only to a specific limit.

The duration of the replication process occupies ≈ 90% of the whole E. coli cell cycle. Proceeding from the corpuscular-wave nature of the given process (see Chapter 3), the mesosome is divided in half during the passage of two replication forks through the mesosome region. A cell growth stops after that as the cell volume, the quantity of protein, carbohydrate, lipid molecules are increased twice. After that the synthesis of biomolecules continues in a fixed volume. It causes the sharp rise of intracellular pressure and a DNA-wave lifts from the 0th to the 1st energy level. The high osmotic pressure and forces of repulsion between two new similarly charged mesosomes lead to a destruction of the cell wall in the mesosome region. They move in the opposite directions and at the same time a double DNA-particle begins to split. The cell volume increases and the intracellular pressure decreases to the primary level. The cell cycle ends with a cytokinesis and a formation of two daughter cells. These new cells are the open systems too and into them the biosynthetic processes and accumulation of substances from outside are continued. That is why they grow rapidly. The excited state of two copies of the DNA is unstable, and DNA-wave are returned to the initial (0th) level, but this process is already realized in the volume of new cells. The emitted quantum of free energy initiates the next cell division and the list of intracellular events proceeds in the same sequence (Figure 23). From the aforementioned it follows, that the life cycle of E. coli is a highly organized biophysical and biochemical process, which ensures the stability of the work of a biological harmonic oscillator in space and time and depends on the fundamental laws of physics.

Thus, the development of prokaryotic cells from generation to generation proceeds in the same sequence and for the same time period and includes the same events which are connected to the change in the corpuscular-wave characteristics of DNA and intracellular pressure: a DNA-wave radiation, a DNA-particle replication, an increase of intracellular pressure, a DNA-wave translation to the excited state, a nucleoid and cell division, an decrease of intracellular pressure, a DNA-wave radiation, etc. It seems that the process of cell development occurs in a circle. Here it is important to remember that a DNA-wave is the standing wave. The energy enters into a cell by the two streams and initiates the development of two equal cells which cross over to live in the new living space at the end of every cell cycle. That is, development processes are not returned to the starting point but continue on to the two new ones. Therefore, the principle of development of prokaryotic cells is not a circle. It is a spiral (Figure 24).

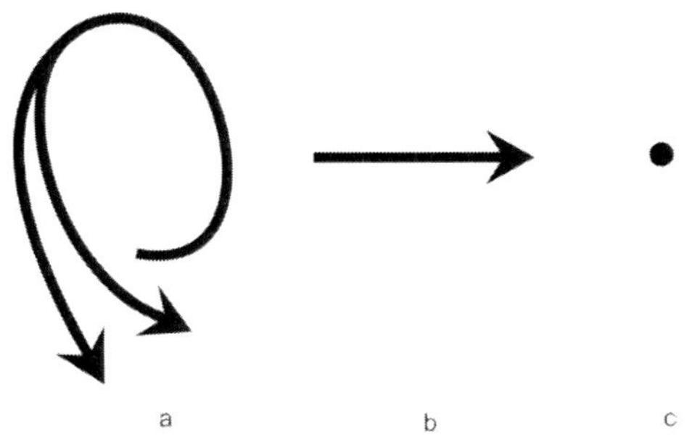

Figure 24. Principle of the development of prokaryotic cell depends on the living state of a cell: a-dividing cell; b – resting cell; c – dead cell.

Spore and Resting Cell

A sporulation (a process of the formation of endospores or cysts) is one of the most ancient ways of survival for prokaryotes in an unfavorable environment. In Cyanophyta, Bacillus, Clostridium, Sporolactobacillus, Sporosarcina and etc. genus of bacteria form spores of a special type – the endospores. It develops inside a sporogenous ("maternal") cell which becomes a carrier, protection structure and source of nutrients at all stages of the sporulation processes. The sporulation is a complicated multi-stage process during which the proliferous prokaryotic cell passes into the resting state. The spores remain viable for a long time and can endure the high and low temperature, drying, influence of radiation, acids and poisons, which are lethal for normal vegetative cells.

The sporulation as a biological process that can be divided into three stages on which the leading events are: 1) a change of the morphology of the initial sporogenous cell (the fusion of nuclear substance along a long axis of maternal cell); 2) a prospore arising and its development (the isolation of a nucleoid zone with a thin layer of cytoplasm and the invagination of the plasmatic membrane; the prospore separation from the "maternal" cell by means of a two-layer plasmatic membrane); 3) a spore mating (the formation of the layers of sporous coverings; the destruction of a "maternal" cell and its external envelope and the exit of spores from the "maternal" cells).

The cells during the transition in a resting condition decrease in size. Then the division of the cell is stopped. The sporulation beginning is connected with the active synthesis of proteins, RNA and others biomolecules. Its end is accompanied by a decrease of metabolic processes, with the accumulation of supply substances which are deposited into a cytoplasm in the form of granules. All spores have a stable envelope. The connection with an environment is almost completely absent. Therefore a spore is a closed system. The rest state is a reversible condition. When a spore reaches favorable conditions it swells, germinates and begins to divide. We have attempted to explain the mechanism of sporulation in bacterial cells proceeding from the corpuscular-wave nature of this process, by the experimental data enumerated above and the conditions of cultivation. The peculiarities of the development of bacteria in a higher temperature regime (35-40^0 C) were analyzed for this purpose. In this form we must remember that the viscosity of majority liquids (including the protoplasm) is diminished with a temperature increase.

The formation of future sporogenous cells in conditions of higher temperature begins also from a DNA-wave radiation and a doubling of DNA-particles. Then the two-ring chromosomes, which is associated with the cell membrane by means of mesosome, begins to move to opposite sides, as far as the growth of cell. In this time the DNA molecule presents 1 DNA-particle and 1 DNA-wave. The DNA-wave is situated on the 1st energy level. In fact these chromosome-waves are the two synchronized vibrators. They as vibrators have the same frequency and amplitude of the oscillations. The perturbations, which are generated by two oscillating chromosomes into the less viscous protoplasm (in conditions of higher temperature), do not remain localized. Therefore the probability of an arrival into a cell of a common oscillating contour on the basis of two DNA-particles and a thin protoplasm layer is sufficiently greater. This wave is a short-duration process and circulates along the long axis of the cell. The oscillating contour is destroyed as the result of a removal of chromosomes from each other during a cell growth cycle. Cytologically this phenomenon could be registered, as an unusual state of the nuclear substance. During this time the synthesis of protein, RNA, lipids, etc. are continued.

In figure 25 the change of the energy state of DNA-wave in bacterial cells at a normal development and at a sporulation is shown. We propose that the physical state of sporogenous cells should be satisfied by the two main requirements for the beginning of the formation of a cell-spore: 1) high intracellular pressure and 2) the coincidence of a longitudinal axis of a cell in the direction of gravity. In this case the transfer of the energy from chromosome to chromosome without DNA-wave radiation can be realized, in accordance with the law of energy conservation (by analogy with a 2nd meiotic division during activated meiosis, chapter 3). I.e., the DNA-wave of a chromosome from the apical pole loses a quantum of energy and is transferred from the 1st to the 0th energy level (nucleoid of maternal cell) (Figure 25, on the left). The DNA-wave of a chromosome from the basal pole receives this quantum of energy and is lifted from the 1st to the 2nd energy level (nucleoid of spore). Accordingly, the nucleoid from the apical zone gives rise to the endospore. The nucleoid from the basal zone remains in the "maternal" cell, which eventually dies.

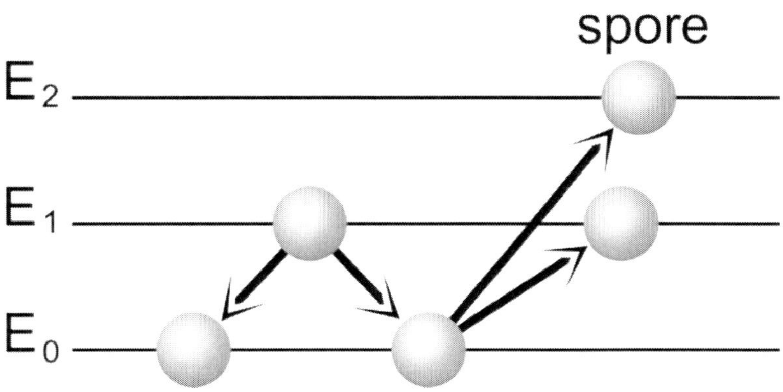

Figure 25. The scheme of energy motion in bacterial cells with a sporulation and during normal development.

After that inside a sporogenous cell a situation occurs when the two biological harmonic oscillators with different capacities begin to work at the same time. They create a biofield which is distinguished by the intensity and radius of action. As known, the most the biomolecules (proteins, nucleic acids, lipids, mono-, di- and polysaccharides) are the molecular dipoles. Therefore the ions and biomolecules are set in motion and begin to concentrate around a more powerful source. The centripetal movement of lipids leads to the plasmatic membrane rise, the prospore emergence and its separation from the "maternal" cell. It should be noted that the mechanism of the formation of a plasmatic membrane is similar to the process of nuclear envelope restoration after mitosis (chapter 2). Then the period of spore maturation and formation of firm and stable cell wall begins. By a consequence the growth of molecule concentration in a fixed volume of spore is: an increase of intracellular pressure, diminishing of water content, decrease of metabolic activity and a synthesis of biologically inert substances. At the time of spore maturation the DNA-wave of the "maternal" cell (the 0th energy level) breaks down and collapses due to friction.

A biological harmonic oscillator as the physical system can work in a specific (effective) zone of pressure and temperature only (chapter 1). Therefore for any cell system exists in the upper threshold of a pressure the narrow zone in which limits the action of the oscillator is stopped, but a DNA-wave is not destroyed. In this zone the spore biofield disappear; the cell becomes a closed system and enters the resting state or in the equilibrium at the critical point. A bacterial cell can leave the resting state only after activation of cell system by means of external forces. A spore after transference in a fresh nutrient medium and at an optimal temperature regime begins to absorb water and swells. During swelling the cell volume is increased, intracellular pressure is down and the wave function of the DNA is resumed. A DNA-wave passes to the 1st energy level and emits a quantum of free energy. The cell wall due to this energy is broken and the spore begins to germinate. Not long after the DNA-wave lowers to the 0th level, a bacterial cell renews the process of replication and division.

Thus a spore represents a living system, whose development proceeds at a very low velocity. All vital processes are not destroyed and the state of rest is reversible. The main metabolic reactions are not stopped completely and the journey in this state is not infinite. Therefore the principle of the development of a spore as a resting cell is a line (Figure 26).

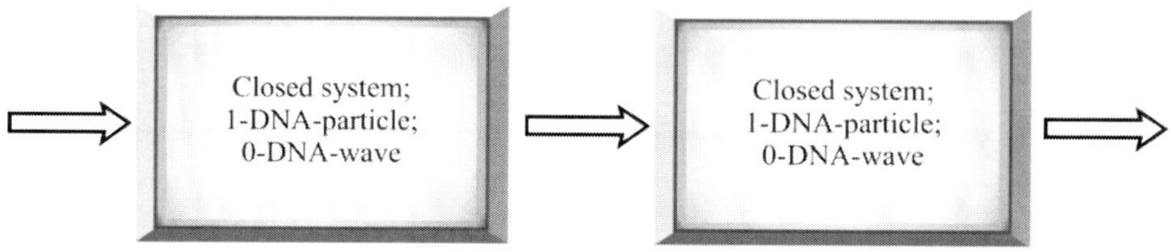

Figure 26. The principle of the development of resting cell or spore is a line.

Dead Cell

The death of prokaryotic organisms under natural conditions takes place often and occurs for different reasons. A deep abyss separates the living and dead cells. During its death the DNA-wave disappears, intracellular homeostasis is interrupted, cell contents are destroyed, the DNA-particle is fragmented and the plasmatic membrane and cell envelope have lost their integrity. Therefore, the principle of the development of dead cell is a point, or the absolute lack of development (Figure 24).

We can assume that the transition from living cell to dead cell is connected with the change of its wave properties. A cell death becomes an irreversible process after the destruction of the wave function in the DNA or the breach of the structural integrity of the biological harmonic oscillator. Indeed, the first consequence of their degradation is the disappearance of the biofield. Later the event is occurring in the following ways. Without a biofield the concentration of the ions in the cells is changed, all energy, synthetic and metabolic processes, including a DNA synthesis, are broken and the cell itself swells.

The data testifies to the fact that the whole life of the prokaryotic organisms must be directed to ensure uninterrupted operation of the biological harmonic oscillator, as the source of inner energy. The stability of its work is achieved mainly through the homeostasis of intracellular processes under the conditions of a constantly changing environment. For this reason, the speed of the synthetic processes, the dynamics of intracellular pressure and the time of DNA-wave excitation - radiation during the cell cycle are strictly coordinated with each other, but only up to a specific point of variability of the influences on it (temperature, osmotic pressure, thickness of cell walls, etc.). Any breach of cell metabolism or other powerful stress, which is caused by internal or external reasons, leads to an imbalance of the whole intracellular system, a damping of the wave function on the DNA, a stop of the work of the biological harmonic oscillator, a beginning of necrotic processes and a cell death.

Eukaryotic Cell

The eukaryotic cell is considered to be a highly organized type of cell system. Its dimensions on an average are much larger than that of prokaryotes. The size and form of the plant cell is strongly varied in its dependence on function and position, which it occupies in a plant. A cell has a more complicated morphology as well. The presence of a nucleus is the distinctive sign, which consists of a nuclear envelope, chromosomes, nucleolus and a karyoplasm. The main evolutionary properties of eukaryotic cells in comparison with "prenucleus" organisms are as follows. 1) After the appearance of a nucleus the biological harmonic oscillator is displaced in its volume (or inside a new closed cavity with reflecting walls). 2. The cell cycle of proliferous cell proceeds in the two systems – a closed and an open. 3. The whole cycle of the individual development of multicellular organisms is characterized by the alternation of generations (a haploid and a diploid). 4. A fixed mesosome is substituted on the mobile analogue (two centrioles). 5. A second source of internal energy is created (metaphase chromosomes). 6. A processes of replication and transcription are separate in time and in space. 7. An increase of the volume of the nuclear genome takes place. The numbers of chromosomes become greater, the chromatin structure more complicated, the main components of which are DNA, histone and non-histone proteins. 8. A doubling of the DNA-particles is a multistage process with a great number of replicons. 9. An interphase chromosomes are connected with a nuclear envelope at three points - in the centromere and two telomere regions. 10. A cytoplasm consists of the organelles (mitochondria, chloroplasts, endoplasmic reticulum, Golgi apparatus, etc.), inclusions (tannins, starch, lipid globules, etc.) and a hyaloplasm. 11. A mechanisms of osmoregulation are determined in accordance with the ascending (sexual development) and descending (ontogeny) streams of energy during the individual life of a multicellular organism. The homeostasis of intracellular pressure is a necessary condition for a unidirectional movement of biological energy during the ontogeny of cells and plants at whole. 12. The universal scheme of mitosis, meiosis, and reproductive cycle for each species and mechanism of change in the DNA-particle (cross-pollination, crossing-over) and the DNA-wave (asymmetric mitosis, activates mitosis) are created.

Then we began looking among the enumerated signs for the leading link, the change of which entails all the rest of the evolutionary transformations. In this analysis the degree of the evolutionary perfection of structures which is responsible for the energy supply of a cell is, apparently, the main and we used it as the criteria for the consideration of the processes of progressive evolution.

As known, all living cells spend energy with a very high efficacy. A cell is characterized by a complicated corpuscular-wave organization, many elements of which progressively developed and improved during evolution - from the simple to the more complex. A cell system for the creation and maintenance of this complexity and order requires a high level of energy supply. Therefore the improvement of existing and origin of new energy sources and channels for energy utilization must be precede the beginning of any evolutionary process. Without the additional energy is starts neither a single biological reaction nor a single structural transformation.

The energetics of the main part of prokaryotic organisms is provided by the energy from a biological harmonic oscillator and the energy from different types of the external sources. The more progressive species (blue-green algae, photosynthetic bacterium) produce the energy for metabolic processes by means of visible light. The fermentation is the most widespread way among prokaryotes for the extraction of energy from a nutrient and is characterized by low efficiency. The energy capacity of a biological harmonic oscillator corresponds to the low limit of the action of an oscillating system: the base state is the zero energy level and the excited level is the 1hω energy level. The bacterial-type of phosphorylation is at an early stage of the development of mechanism of the photosynthesis.

The energetics of the eukaryotic cell demonstrates a high differentiated and powerful system for obtaining and using free energy. In plants it is comprised of four independent sources – 2 internal (DNA-wave and metaphase chromosomes) and 2 external (respiration and photosynthesis). During evolution the energy potential of a cell is essentially increased due to the improvement of the design of existing sources and the origin of one more source – metaphase chromosomes. Eukaryotic cells have more effective ways of energy acquisition in the system "ATP - ADP", which proceed on the basis of the processes of oxidative phosphorylation and photosynthetic phosphorylation. A cytoplasm has special organelles, the functions of which are related to the exchange of energy. All reactions of photosynthesis take place in the chloroplasts. The processes of oxidative phosphorylation, that are associated with the Krebs cycle, occur in the mitochondria. A biological harmonic oscillator is enclosed in a volume of the nucleus. Such synthetic processes as replication and transcription are provided by the energy from different internal sources. During the cell cycle the joint work of internal and external sources fully maintains the viability of the cell in wide diapason of environmental conditions. Certainly, all these energy sources have great value in a cell life. Nevertheless, we consider that the biological harmonic oscillator among them occupies a particular significance for the following reason. Without cell division, life as we know it does not exist. The energy of this source not only initiates a process of the formation of new cells, but also creates the mechanisms for its realization (spiralization of chromosomes, formation of centrioles, etc.). This and other questions, connected with the leading role of a biological harmonic oscillator will be discussed in this chapter in more detail.

Apparently, the transference of a biological harmonic oscillator in a nucleus is the main event of cell evolution, for the reason that the reliability of the oscillating system became higher. Simultaneously the quality (Q) and energy potential of the oscillator is increased, because, first, a chromosome-wave, which is oscillating within the nucleus, is protected better from the internal (organelles) and external environment and, second, a loss of friction and the velocity of wave decay are lower. Thanks to this transformation cells begin to "master" higher energy levels (n = 2hω, 3hω, 4hω, etc.), and living matter can cross from the unicellular to multicellular level of organization.

The duration of the cell life on a basis of one DNA (particle and wave) is lengthened considerably due to the increasing the number of living states, which the cell may have during its own individual development and the position on plant body. In high order plants there are five such cell states: proliferative, differentiated, equilibrium, resting and dead cells. Each state corresponds to its own principle of the development of a cell system. In the description of the living states of eukaryotic cells were used in a number of reviews on a cell biology and the developmental biology of animals and plants (Gurvich, 1932, 1944; Brachet, 1957, 1960; Mazia, 1961; Lehninger, 1965, 1985; Margulis, 1970; Clowes, 1971; The cell cycle, 1973; Life of plant, 1974; Esau, 1977; Swanson, and Webster, 1977; Singh, 1978; Bell, 1980; Cantor, C.R. and Schimmel, 1980; Danilova and Kashina, 1980; Kozubov et al., 1982; Volkenshtein, M.V., 1983; Tokin, 1987; Merkis, 1990; Alberts et al., 2002; Rubin, A.B., 1999; Rees, and Sternberg, 2002; Zhimulev, 2003; Chentsov, 2004; Belousov, 2005; Gilbert, 2010; etc.).

Proliferous Cell

The cell cycle of the proliferous cell of plants has been described in detail in Chapter 2. Therefore we discuss only its main processes, and the energy positions. Figure 27 shows the scheme of the cell cycle of a plant cell between the two successive radiations of DNA-wave. The initial position of a cell genome is 1 DNA-particle and 1 DNA-wave. According to our notions the biological harmonic oscillator as an oscillating system is in action during interphase, but as a source of internal energy it works only once a cycle – somewhere in end of the G_1-period. The cell is an open system. In this time a cell is exchanged by energy and matter with the environment and all sources take part in its energetics – the two internal (DNA-wave and metaphase chromosomes) and the two external sources (photosynthesis and respiration. The external energy comes into the cell in the form of different nutrients and light quanta, which are caught in molecules of chlorophyll. This energy during the process of corresponding complex reactions is converted into the energy of ATP chemical bonds. The supply and utilization of energy in the system of ATP-ADP is continuous. As a consequence, the total volume of the energy balance of an interphase cell is maintained at a constant level.

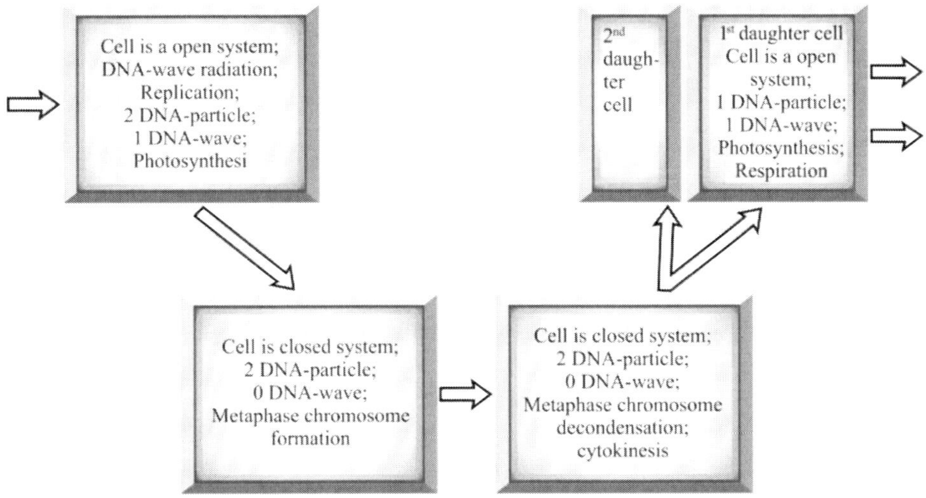

Figure 27. The scheme of the development of proliferous eukaryotic cell of plants.

A cell doubling initiates the quantum of free energy, which enters into a nucleus during the transition of the DNA-wave to a lower energy level. A largest part of this energy is spent on the DNA-particle biosynthesis, chromosome condensation and centrioles creation. The rest of energy is stored in the ATP chemical bonds, and is used for a cell growth and is dissipated as heat. Afterwards a DNA doubling genetic apparatus begins to exist in the form of 2 DNA-particles and 1 DNA-wave.

The formation of the 2nd internal source of energy begins from the end of the S-period and is lasts through the entire G_2 period, prophase and metaphase of mitosis. The condensation of chromosomes takes place by means of two different mechanisms - the inner (G_2-period) and external (prophase and metaphase of mitosis) coiling. The balance of chromosomes during inner coiling is altered as result of the one-sided turns of the nucleotides (as the isomeric links of a DNA molecule). The DNA-particle during this process is gradually shortened, and its internal energy is increased. At the moment, when the length of chromosomes becomes shorter than the diameter of nucleus, they separate from the envelope of nucleus. The DNA-wave, biofield, and nucleus envelope are destroyed. Respiration, photosynthesis and all synthetic reactions are stopped. The cell becomes a closed system and is characterized by a maximum level of entropy. Another part of the intracellular events will be implemented as a physical process into a closed system.

Two positively charged centrioles (which during G_2 period by means of the forces of the biofield were kept inside a nucleus) are repulsed reciprocally after the biofield disappearance and move in different directions, giving rise to the formation of two cell centers. During the centrioles divergence the doubled chromosomes are split at the chromatids up to their own centromeres. After that the genomes exist as 2 DNA-particles and 0 DNA-waves. Between the centrioles, as between the two positively charged organelles, an electric field is created. A chromosome set, caught in the zone of the action of the electric field, renews the process of condensation by means of external coiling. The chromosomes are going out in the area of the equatorial plate, where the maximum degree of condensation is reached. They, as super compressed "springs", contain large amounts of elastic energy. The size, shape and position of them on the equatorial plate remain the same, and the mechanism of the metaphase chromosome formation, as a source of internal energy, is a very orderly process. It is important to emphasize, that the carrying over of the last stage of condensation in the condition of a closed system significantly increases the preciseness of the packing of the chromatid coils, due to the decrease of different hindrances, which are the result of metabolic and functional activity of the cell in the conditions of an open system. Due to the high level of the organization of the coiling process, each chromatid up to the metaphase end has an equal amount of elastic energy.

The chromatids during anaphase are separated from each other. At this time a cell is a closed system too. After the separation of chromosomes the DNA molecule is represented by 1 DNA-particle and 0 DNA-waves. The two groups of chromosomes move in different directions under the action of the forces of Coulomb attraction between the positively charged centrioles (or the two molecules of a polymeric actin) and negatively charged centromeres. At the time of contact their charges are neutralized and the molecules of the polymeric actin are split based on the initial number of the monomers of actin, as a consequence the centrioles cease to exist. The electric field after neutralization charges disappears and the chromosomes begin to untwisted and come back to the initial (interphase) electrochemical structure of the DNA. The energy, which is extracted at the time of decondensation chromosomes will enter the newly formed daughter cells and causes their energy level to rise considerably. The ATP pool is increased and the enzyme of the RNA-polymerase is activated.

Thus, a telophase is the sole stage of mitosis, when a cell is supplied with free energy from a powerful source, which was created for this objective the DNA-wave and special intracellular mechanisms during the G_2 period and mitosis (prophase and metaphase). So much more energy enters the cell when it is a closed system, which is what considerably raises its efficiency. In a closed system this energy is stored by means of transformation into other forms for performing useful work after division. So the newly formed daughter cells always receive the same amount of energy, which is not changed or dependent on organism age or environmental conditions. That is why the phenomenon of the "metaphase chromosome", as the independent internal source of energy, has a specific place in the life of the eukaryotic cells. The analysis of the experimental data has shown, that the energy of the 2nd source is primarily used on the creation of the two energy streams - a centripetal (the ATP-ADP system) and a centrifugal (the processing), a cell growth, an excitement of biomolecules and a cell transfer from a closed system into an open system.

Interphase is the longest stage of the cell cycle. It exists during a G_1-period, S-period and G_2-period. At the G_1-period, as a result of the DNA-wave and biofield restoration, a cell begins an active exchange with the environment and becomes an open system. Its genetic apparatus is returned to its initial state of - 1 DNA-particle and 1 DNA-wave. A cell during the G_1-period is growing rapidly at the expense of the kinetic energy received from the 2nd internal source. The ultrastructure of the cytoplasm is reformed, according to the corpuscular-wave characteristics of its DNA-particle and DNA-waves. It is characterized by a high level of categorization and orderliness. The membrane structures (chloroplasts, mitochondria, endoplasmic reticulum, vacuoles, Golgi apparatus, etc.) are the means for a localization of cytoplasmic sections, wherein is created all the proper conditions for the realization of strict and specific enzymatic reactions. At the same time the different metabolic processes which are associated with reactions of anabolism and catabolism are renewed and intensified. The activated enzymes of the RNA-polymerase are attracted to one DNA thread and form with it a durable bond. Then each enzyme molecule during movement together with a DNA-wave begins to unwind the double spiral of a DNA-particle and simultaneously carries out the biosynthesis of the RNA of the new generation (mRNA, tRNA, and rRNA). As consequence, the centrifugal and centripetal energy streams which function on the basis of processing and ATP molecules begin to operate.

The channels through which energy is moved during the cell cycle are an important part of energetics of the cell. We may suppose, that the centripetal and centrifugal energy streams have existed in all interphase cells. They differ by origin, nature and the means of utilization. Both these streams originate from the 2nd internal energy source and the time of their activity is limited by the interphase only. The centrifugal stream operates on the basis of the energy, which is extracted during the processing. The centripetal stream works in the system ATP-ADP only. This processing by definition is a multi-step orderly process of maturation of RNA, when the large molecules of the primary RNA-transcript are converted to the mature RNA. Its mechanism according to the manifestation and behavior during cell cycle very often resemble such physical phenomenon as an exponential decay. This process often occurs in nature and is accompanied by the acquisition of free energy. Actually, a decay of RNA molecules is constantly taking place in the nucleus, nucleolus, ribosomes, and cytoplasm, and it is the main characteristic of the processing phenomenon. The functional purpose of the centrifugal stream is due to a securing of energy for protein biosynthesis. Therefore in our model the process of translation involves all types of RNA: an mRNA serves as the matrix, a tRNA is required for code reading, and an rRNA is the energy reservoir. Their joint work is designed so that the quantity of protein is always doubled at the end of every interphase. The maximum intensity of the centrifugal stream is observed at the beginning of the interphase, and the minimum level at its end, when the largest part of initial molecules should be in the form of mature RNA.

Another energy channel, which acts in the opposite direction, is the centripetal stream. The vector of its action is directed from the external environment into the cell. It operates in the ATP-ADP system and its work in plant cells draws on all four sources energy - a DNA-wave, metaphase chromosomes, photosynthesis and respiration. A 2nd internal source creates this stream due to the energy of a metaphase chromosome. The two external sources together with a DNA-wave replenish the pool of ATP during interphase, preventing, thus, a rapid decrease in its level in the cell. By the interphase end the energy of the two streams is weakened. Therefore, the only way for a cell to survive is to enter into a new cycle of cell reproduction. The development of the cells of the next generation begins from the DNA-wave radiation and continues up to the moment of irreversibility, when the division in a cell life is not possible to abolish.

The genome of the eukaryotic cell is characterized by the recurrence of development. As well as a prokaryotic cell in its basic state is 1 DNA-particle – 1 DNA-wave. A DNA-wave radiation is initiated during the process of replication and the first change of the genome. The end result is also the formation of two new cells with an identical DNA-particle and an identical DNA-wave. However in eukaryotes the full turn of a cycle includes the four types of changes in the corpuscular-wave characteristics of a macromolecule (1 DNA-particle – 1 DNA-wave → 2 DNA-particle - 1 DNA-wave → 2 DNA-particle - 0 DNA-wave → 1 DNA-particle - 0 DNA-wave), while in prokaryotes it consists only of two types (1 DNA-particle – 1 DNA-wave and 2 DNA-particle - 1 DNA-wave). That is why it seems, that the prokaryotic and eukaryotic cells have the same principle of development – the principle of a spiral. For the confirmation or refutation of this statement let's analyze the principle of development of a eukaryotic cell during cell cycle once more.

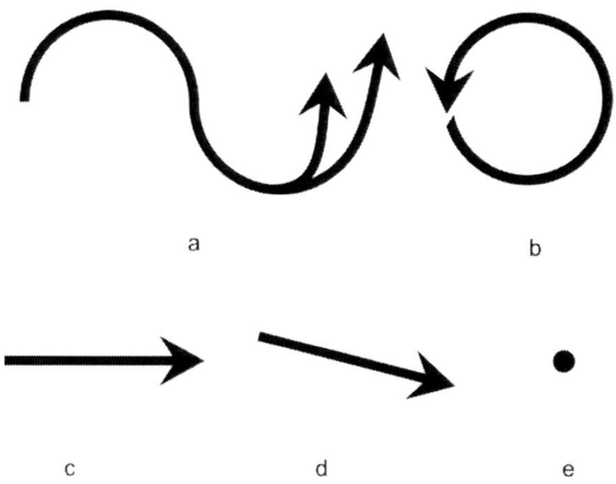

Figure 28. Principle of the development of eukaryotic cell in depends of living state of cell: a- proliferous cell; b – differentiated cell; c – resting cell; d – equilibrium cell; e – dead cell.

Indeed, the basic processes of the genesis of new cells within the mother cell begin to unfold at first in a circle. However, in eukaryotes the mechanism of mitosis interrupts the action of the biological harmonic oscillator for a period of the cell division, and in prokaryotes the mechanism of promitosis does not. The destruction of the oscillator stops the development of the genome by the principle of a circle and, accordingly, of a spiral, and switches it on the wave principle of the organization of the process. The period of the mitotic development of the genome (the lower half of the wave) is a purely one-plane (corpuscular) process, since a DNA-wave at this time is absent, and the given biophysical process proceeds inside the cell, which is a closed system. In spite of that the life of the maternal cell at this time is ending (the metabolism and internal structure are destroyed) and the genetic material of future cells is preserved in full. The mitotic development of the genome inside a closed system ensures the preciseness of its passage through the mitosis phases, the equal distribution of genetic material and energy between the daughter cells, the simultaneous motion of chromosomes up to the centers of formation of new cells, etc. The work of the 2nd internal source of energy reestablishes and "switches on" the biological harmonic oscillator, wave function of the DNA and energy potential of the cell and also separates the two daughter cells from each other. The new cells become the open system and begin the stage of ontogenesis and continue to develop during interphase in accordance with their own wave (the upper half of a wave). Therefore, the principle of the development of a eukaryotic proliferating cell - is a wave (Figure 28).

Differentiated Cell

As known, the whole life cycle of multicellular organisms is divided into the ontogenesis and sexual reproduction periods. During the ontogenesis period is realized a corpuscular-wave program of the development of a plant, which was placed into the last generative cell (zygote, meiospore, etc.) on the basis of unique DNA-particles and a DNA-wave. As the DNA-particle line: the cells of any multicellular organism obtain the identical DNA-particles according to the principle (the similar originates the similar), or in the modern interpretation of cell theory - all cells of one organism contain the same genetic information, which is passed from cell to cell during cell divisions. This conclusion is now confirmed in a large amount of experimental material. As in the DNA-wave line, the cells of any multicellular organism are developed by the descending vector of energy motion, or in accordance with the principles of progressive "damping" of the DNA-wave. The "damping" of a DNA-wave is a genetically determined process. It is based on the action of a quantum-mechanical system – a biological harmonic oscillator, whose work is entirely governed by the fundamental laws of physics, including the laws of thermodynamics. The life of a new organism begins with the first division of an initial generative cell or nucleus (zygote meiospore, etc.) and harmonic oscillator of which has a maximum of biological energy. During ontogenesis the amount of biological energy is varied in a one way direction only - from the maximum (the species-specific level of a zygote) to the zero (the level of the natural death of individuals). This one way direction of the work of a biological harmonic oscillator is provided by means of the conditions of physiological homeostasis, which is created in the all cells and remains at a constant level during the entire life of the plant.

Thus, a biological harmonic oscillator as the physical system, starting from the first zygote division up to a plant death, with definite regularity comes down from level to level in accordance of the second law of thermodynamics. The first generations of cells, as a rule, is a totipotent and has the ultrastructure of a meristematic cell. The decrease of the energy potential of a cell modifies its structural organization, metabolic and functional activity. This leads to the loss of totipotency, to the processes of differentiation, aging and death of a cell. It should be noted that the energy state of genome during ontogenesis is accomplished with the participation of the two types of cell divisions – usual and asymmetric mitosis. Therefore, at the same time in a plant body there are cells with a different degree of differentiation, namely a meristematic, feebly differentiated, highly differentiated, old and dead cells.

The eukaryotic cell for maintaining of metabolic activity, intracellular homeostasis and ordering of structure in a metastable state requires a large amount of energy. The continuous fall of its levels during ontogenesis is a result of "damping" of a DNA-wave and leads to a gradual decrease of biofield intensity inside and outside of a cell. An aging cell for restraint of entropy growth and preservation of viability will require more and more energy. It is known that the replication, transcription and mitosis are the most energetically expensive processes of the cell cycle. A large part of the energy from the 1st internal source is utilized on the DNA biosynthesis, and the energy of 2nd internal source is consumed on the RNA biosynthesis. That is why a cell from a specific moment is not able to ensure by energy alone these two processes and at the same time maintain the viability and produce the energy. The only chance for the survival of an aging cell is the transition of the cell cycle to a more economical means of development. This is achieved by the reorganization of its structure, and energy cannels, namely the stage of replication is removed from the cell cycle and the energy of 1st internal source is switched on in the transcription process. In this case, first, replication and mitosis have disappeared as the two most energetically expensive processes and, second, the cell life on the basis of one DNA molecule (a particle and a wave) is lengthened. The energy, which is liberated during this transformation, will go on to perform the maintenance of a cell metabolism. As a result a cell ceases to divide and passes on in the category of a differentiated cell.

A differentiated cell is an open system. Its genetic system presents as 1 DNA-particle and 1 DNA-wave and is not changed until a cell life comes to an end. A cell cycle becomes shorter and now the cycle is bound between the two consecutive radiations of DNA-wave. A biological harmonic oscillator operates during a cell cycle and through the entire life of the cell up to its death. The number of energy sources is reduced to three (biological harmonic oscillator, photosynthesis and respiration). The energetics of a cell during a cell cycle is provided by the two contrary streams of energy - a centrifugal (based on processing) and centripetal (based on ATP-ADP system). In differentiated cells a great part of the energy from internal sources is spent on the activation of the enzyme of the RNA-polymerase and on the formation of ATP molecules, i.e., on the creation of centripetal and centrifugal streams of energy. Once again we will accentuate the fact, that the transition of a proliferous cell to an irreversible differentiation is connected to the loss of one of the energy sources (metaphase chromosomes).

A corpuscular-wave model of development of the differentiated cells is shown in the scheme (Figure 29). A new cell cycle initiates a DNA-wave radiation and entry into the cell of the quanta of free energy. After activation the RNA-polymerase molecules are attracted to one of the threads of DNA. Every enzyme molecule unwinds locally into two threads of

DNA, when it is travelling with its wave, and simultaneously carries out the biosynthesis of the large molecules of RNA to the next generation (pre-rRNA, pre-mRNA, pre-tRNA). At this time the centrifugal stream is replenished of its energy due to a great number of synthesized molecules of the precursor of RNA. As a consequence, the energy balance of the stream is restored. The centrifugal stream is formed once in the cycle (after a DNA-wave radiation), and then its intensity decreases as far as an exponential decay of RNA molecules. The energy which is liberated during processing is spent mainly on a biosynthesis of proteins. Three sources take part in the creation of a centripetal stream – the internal (DNA-wave) and two external sources (respiration and photosynthesis). After DNA-wave radiation the pool of ATP is raised. Then during the cell cycle its levels will be gradually decreased, despite the fact that this stream is continuously replenished with ATP molecules, which are formed during photosynthesis and respiration. Naturally, it is necessary to provide large amounts of energy for cell life. A differentiated cell uses this energy for different purposes: an execution of work, an energy supply for different synthetic processes, a creation of orderliness from a disordered state, etc. That is why the intensity of the centripetal and centrifugal streams reaches a critical level at the end of every cell cycle. In accordance with the energy potential the entropy level will be increasing as well. The next DNA-wave radiation will raise the energy potential of the differentiated cells and initiates the intracellular events in the same sequence.

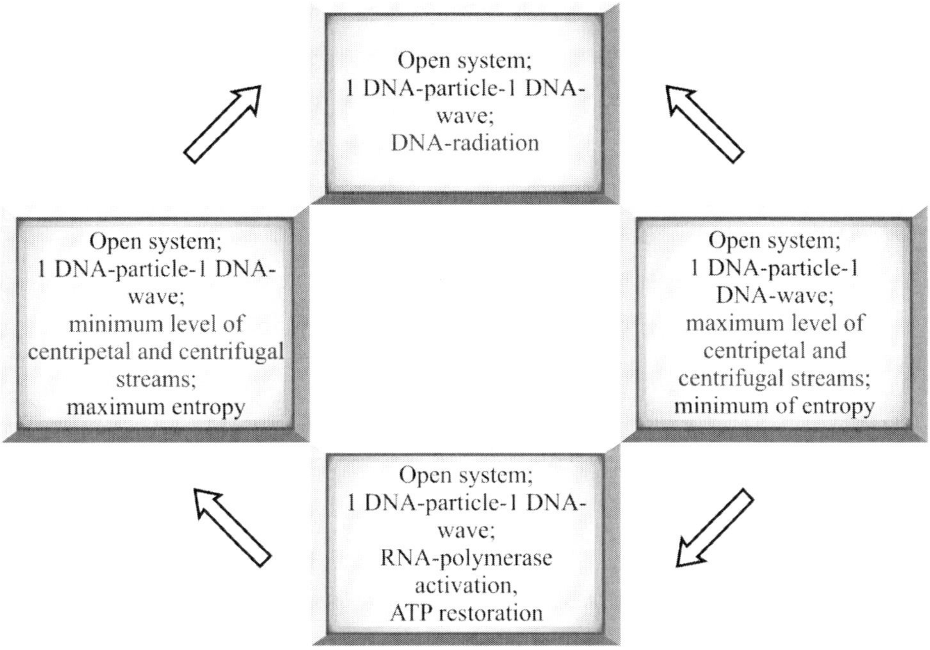

Figure 29. The scheme of the cell cycle of a differentiated cell.

The constant decrease of energy potential, as a result of DNA-wave transition to lower energy levels, is going on during the entire life of the cell and the organism as a whole. Cells, which have entered on a path of differentiation, are changing definitely by the morphological, physiological and functional characteristics. The meristem during differentiation is transformed in the different specialized tissues (parenchima, cambium, etc.) and loses the property of totipotency. Before the stable differentiation will be reached and a cell ceases its existence the divisions should be replaced by several generations. And even during the stage of cytodifferentiation (vacuolation, decrease of organelle number, etc.) the cell development continues and goes in the direction of a more pronounced specialization and aging. The given phenomenon may have only one explanation. A DNA-wave, in accordance with the second law of thermodynamics, is descended with the identical periodicity from level to level both in the proliferous cells and in the differentiated cells.

The problems of the individual development of plants are numerous and complex and we will not perform the task of analyzing all its aspects. Let us consider therefore only a few questions of the differentiation and specialization of cells. As it has been shown the behavior of the cytoplasm and nucleus is different during differentiation. The nucleus (a size, form and number of chromosomes) does not change essentially because, as the physical system, a biological harmonic oscillator should have the same structure as a mechanism. On the contrary cytoplasm, the ultrastructure of which is determined by the condition of a biofield, undergoes powerful changes. The derivatives of the meristematic cell after one or more divisions begin to vacuolizate, increase in size, specialize, cease to divide and eventually die. Moreover, each cell line on the path of differentiation has its own duration.

In the majority of differentiated cells there are the specialized cells in the structural or functional directions, which fulfill one or more specific functions. We can assume that a weakly differentiated cell during transition (from proliferation to irreversible differentiation) and during specialization period has a surplus quantity of energy, which is retained in the cell after a DNA-radiation and with the absence of mitosis. Therefore a majority of them at this stage of differentiation are characterized by intensive growth and usually fulfill a different function. Suspensor cells during lengthening carry the embryonic cells from the ovicell into the endosperm tissue. Note, that the stage of intensive growth in highly differentiated cells is absent. Shown also is that the intensity of the elongation of cells increases in the direction from the young meristematic tissue to a more mature one. Mitotic cell activity is lowered and the rhythm of the work of a biological harmonic oscillator is constant. However a probability of transcription beginning after a DNA-wave radiation becomes higher during cell differentiation. The kinetic energy, accumulating during the cell cycle without mitosis, causes cell growth. The surplus energy may be used for the specialization of cells. For example, the transition from a meristem to a tissue of mesophyll is required in the synthesis of defined enzymes and proteins, which ensure the production of chlorophyll for the photosynthesis. The differential transcription is a distinctive sign of a specialized cell. Perhaps its nature is connected with the wave nature of DNA as well. The activation of the RNA-polymerase in a proliferous cell occurs at the telophase end under the conditions of closed system. The process of transcription begins after a restoration of a biological harmonic oscillator and in that moment a transition of a cell from a closed to an open system occurs. The intracellular conditions, which are created in a cell as a closed system, are similar from cycle to cycle. Therefore the RNA synthesis is carried out at the same sites on the DNA. In a differentiated cell, the transcription proceeds under the conditions of an open system. As a consequence, the matrix for the RNA synthesis can be the different sites of a DNA molecule depending on the position of the cells in the plant body, the condition of the intracellular medium and the external environment. This promotes the development of potential functions and maintains the existence of specialized processes in a cell.

A differentiated cell after the loss of ability to divide, begins to develop within the same living space and is in the same state as the genome (1 DNA particle - 1 DNA-wave). Therefore, the principle of development of differentiated cells is a circle (Figure 27).

Equilibrium Cell

The equilibrium cell is a cell, which has reached the zero energy level during ontogenesis or during a special type of cell division – an activating mitosis. The zero level is the last permissible level of energy, which may have a biological harmonic oscillator as the oscillating system. Below it at the quantum state a macromolecule does not exist. This level demonstrates the equilibrium, the most stable state of any physical and biological system, which is characterized by the lowest energy. Any cell and organism strive for this balance by means of the second law of thermodynamics. The conditions of intracellular homeostasis, which are supported in all eukaryotic cells of plants and animals during their individual life, provide a unidirectional movement of biological energy from a maximum level of up to zero. Moreover, the zero level of a eukaryotic cell unlike a prokaryotic cell is irreversible, because the mechanisms of homeostasis act on all energy levels as well. Therefore a multicellular life should always be finalized by death.

The wave function on the DNA could be acting infinitely, if friction and other forms of energy loss there are absent. In spite of the high quality factor (Q) of a biological harmonic oscillator an energy loss due to friction always takes place. The friction forces an intranuclear medium, which by its physical and chemical nature is a colloidal system, and causes a gradual damping of oscillations on a macromolecule. Therefore, the death of an equilibrium cell is inevitable and is only a matter of time. For some generative cells (prothallial cells, synergids, etc.), the energy of which was lost completely during activating mitosis, the zero energy level is the final part of apoptosis, or a programmed cell death. According to the literature data the duration of their life on this level is different, but no more than 7-10 days. It is established that natural death is often accompanied by a mass cell death, because they were originated on a basis of one DNA molecule (particle and wave) and during ontogenesis many cells have reached the equilibrium state at the same time.

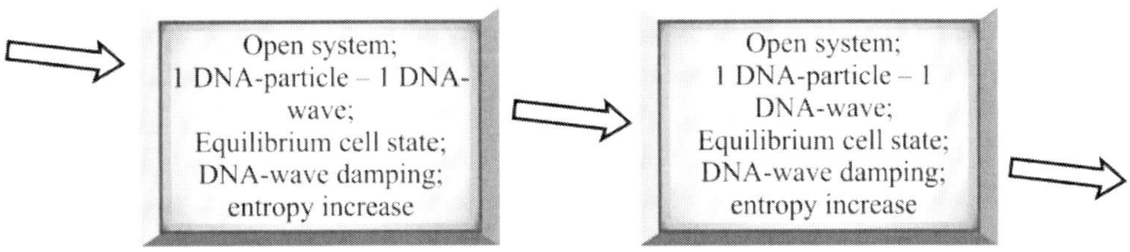

Figure 30. The scheme of the development of equilibrium cell on the zero energy level.

Despite the fact that the equilibrium cell in a series of cell generations is the last link, the reason for its death is not the transition of DNA on the 0th level, but that state in which it has the energetics and metabolism of moribund cell. The life of a eukaryotic cell on this level has its own regularities. It starts from the DNA-wave radiation and ends with the cell death. A biological harmonic oscillator as the internal generator of free energy is manifested for the last time. As the oscillating system and the source of the biofield, it continues to function. The cell is an open system. Therefore, the only external sources of energy for the equilibrium cell are photosynthesis and respiration. The centrifugal energy stream, which was replenished by means of the 1st source of energy for the last time, is gradually disappearing by the reason of processing or the exponential decay of RNA molecules. Accordingly, the quantity of the synthesized proteins becomes less. Centripetal flow is decreased as well, since the external sources cannot cover the energy demands of cell. The equilibrium cell is not able to divide and produce the energy. A complex mechanism of conversion and accumulation of energy on a basis of the reactions of oxidative phosphorylation and photosynthetic phosphorylation is exposed to destruction on account of the degradation of chloroplasts, mitochondria, and the increase of entropy. The necrotic processes embrace a whole cell. The cytoplasm and organelles disintegrate, the activity of the hydrolytic enzymes increases, a biological harmonic oscillator stops, the oscillations of chromosome-waves cease, and a cell passes into the category of a dead cell. Therefore, the principle of the development of an equilibrium cell is a line or more precisely a line going down (Figure 27, 30).

Dead Cell

The death of a eukaryotic cell happens as a result of: 1) natural death; 2) programmed cell death; 3) different cell pathology and 4) random events. During death the destruction of a biological harmonic oscillator always occurs. It is like the "heart" of cell. More precisely it gives a life to a cell, regulates its development and turns it off during death. In the moment of a natural death a harmonic oscillator is gradually discharged in accordance with the second law of thermodynamics, and a cell dies from the old age. For the cell, developing by means of differentiation, it is a "counter" into which is programmed the maximum duration of its life. The oscillator is discharged fully during apoptosis, when one cell gives its energy to another cell during specific cell divisions (activating mitosis, activating meiosis). The cytological picture of its destruction has similar features: a chromatin condensation and fragmentation of the nucleus. A cell life is characterized by orderliness and presents a chain of coordinated events and processes. The violations lead to the different forms of mitosis pathologies, an enzymatic deficiency, an imbalance of the cell system as a whole, a faster discharging of the oscillator, etc., a consequence of which is a breach of metabolism and cell death. If a cell death occurs by reason of external influence (poison, etc.) the biological harmonic oscillator is stopped and is discharged, and the intracellular events are developed as a necrotic process.

The dead cells are part of the life of any living organism of high order plants. In woody plants the greater portion of the body of the plant consists of them, because a majority plant cells have of the thick cell walls. They, as the main structural component of a number of tissues perform the vital functions - protective, mechanical and conductive. For example, the xylem, which is formed from the dead cells with thick walls, is the main water-conductive tissue and a part of the vascular system of plants.

During the death of the cell the cytoplasm and nucleus are destroyed, the cell development ceases and it becomes part of the inanimate matter. Therefore, the principle of the development of dead prokaryotic and eukaryotic cells is the same – a point, or the absence of development (Figure 27).

Conclusion

Thus, our main task in the previous chapters was to ground the hypothesis of the wave nature of DNA and the quantum mechanical nature of a cell and confirm with the experimental data the models of a cell system developed on that basis. These theoretical concepts were further developed in this chapter with an emphasis on the energetics of a cell, its status and the improvement of energy sources and channels of energy transfer within the cell. We have found that the energy processes of a cell are similar in all forms of life. The energy network of a eukaryotic cell consists of the analogous or more perfect type of structural elements as a prokaryotic cell, which includes the external sources of energy (based on fermentation, photosynthesis, respiration), the one internal source of energy (based on DNA-wave radiation) and two channels for the utilization of energy - centrifugal (based on processing) and the centripetal (in the ATP-ADP system). The biological harmonic oscillator performs the leading role. This source of energy is not only a generator of free energy, but also creates a biofield, which creates important benefits for the cells life and its organization of intracellular metabolic processes. The essential distinction between them is the following. In favorable conditions the largest part of the life of prokaryotes occurs at the zero energy level when the cell is in the equilibrium state. The stability of zero level ensures a periodic excitation of DNA and its passage on to the next level (once per cycle). The trigger for the initiation of the new cell division is a returning of a DNA-wave to the initial state. A deterioration of habitat leads to a formation of spores, when the emission of a DNA-wave is stopped, but the corpuscular-wave characteristics of macromolecule are not changed.

The energy potential of a species is progressively increased during evolution mainly as a result of the perfection of the energy network of a cell. It is shown that all origins of a prokaryotic cell have undergone significant changes. In eukaryotes they became more powerful, specialized and their functioning is characterized by a greater efficiency. The additional internal source of energy (metaphase chromosomes) appeared in a proliferous eukaryotic cell. After that a cell cycle begins to proceed in the open and closed systems as the corpuscular-wave and one-plane (corpuscular) processes, the reactions of replication and transcription were separated in time and in space. The powerful development of the energy network of a cell system has created a base and given a powerful impulse for more rapid development of the whole organic world.

The quality (Q) and capacity of a biological harmonic oscillator becomes higher after the transition of a DNA-wave inside of a nucleus and the increase of chromosome numbers. It extends the possibilities for the further perfection of the structural and functional organization of eukaryotic organisms. As a consequence a DNA-wave step by step begins to master the higher energy levels ($2h\omega$, $3h\omega$, ... $nh\omega$). This has allowed living matter to pass to the multicellular level and to create many species on this path, improving and developing their organization by means of changing the DNA-particle line, DNA-wave line and both lines together. The life of animals and plants begins to divide during the two periods (sexual reproduction and ontogenesis) in accordance with the programming of the work of the biological harmonic oscillator. The growth of the total biological energy of the species leads to an increase in the degree of the autonomy (independence) of the internal conditions of the living systems from the conditions of the environment. The level of the adaptability of the species to the conditions of its habitat has also become higher.

The programming of the work of the internal and external sources of energy during the cell cycle is of particular importance for the adaptation of the plant to the environment. The internal sources are acting only once in a cycle: the DNA-wave is radiated in the end of the G_1-period and the metaphase chromosomes are untwisted in the telophase of mitosis. The processes of photosynthesis and respiration take place during the entire interphase. The different times that the internal sources are working are in agreement with the duration of the phases of the cell cycle and the duration of a day. At present, mitotic activity of meristematic tissue has been studied in many species of plants. It was established that the first peak of mitotic activity falls in the morning. Therefore the largest part of the interphase is occurring during daylight hours.

The principle of the development of cell systems depends on the level of their evolutionary advancement and the living state in which the prokaryotic and eukaryotic cell resides. The highest level of the organization of a cell cycle has a proliferous eukaryotic cell. It is the principle factor of the wave. Further the cell systems and living states of cells in descending order are arranged as follows: A divisible prokaryotic cell is the spiral. A differentiated eukaryotic cell is the circle. A resting prokaryotic and eukaryotic cell is the line. An equilibrium eukaryotic cell is the line going down. A dead prokaryotic and eukaryotic cell is the point.

References

Alberts, B., Johnson, A., Lewis, J., Raff, M., Roberts, K. and Walter, P. (2002). *Molecular biology of the cell*, 4th edn. Chaffey, N (Ed.), Garland Science, New York.

Allen, J.P. (2008). *Biophysical chemistry*. Wiley-Blackwell Publ.

Bell, L.N. (1980). *Energetics of the photosynthesizing plant cells*. Moscow, Nauka.

Belousov, L.V. (2005). *Bases of general embryology*. Moscow, Moscow State University.

Brachet, J. (1957). *Biochemical cytology*. Academic Press, New York.

Brachet, J. (1960). *The biochemistry of development*. Pergamon Press, London, New York, Paris, Los Angeles.

Cantor, C.R. and Schimmel, P.P. (1980). *Biophysical chemistry*. Freeman, W.H. (Ed.), San Francisco.

Chentsov, Yu.S. (2004). *Introduction to cell biology*. Akademkniga, Moscow.

Danilova, M.F. and Kashina, T.K. (1980). Apical meristem. *Atlas of the ultra structure of plant tissues*. Petrozavodsk, Nauka.

Davis, B.D., Dubecco, R., Eisen, N.N. and Ginsberg, H.S. (1990). *Microbiology*. 4th edn. Lippicott Comp., Philadelphia.

Esau, K. (1977). *Anatomy of seed plants*. Wiley J. and Sons, Inc., New York, Santa Barbara, London, Sydney, Toronto.

Freifelder, D. (1983). *Molecular biology – a comprehensive introduction to prokaryotes and eukaryotes*. Science Books International, Portola Valley, California.

Gilbert, S.F. (2010). *Developmental biology*. 9th ed., Sinauer Associates, Inc., Sunderland, Massachusetts.

Gurvich, A.G. (1932). *Mitogenetic radiation*. Medgiz, Moscow.

Gurvich, A.G. (1944). *Theory of biological field*. Sovetskay nauka, Moscow.

Kozubov, G.M., Trenin, V.V., Tihova, M.A. and Kondratieva, V.P.(1982). *Reproductive structures of gymnospermae (comparative description)*. Leningrad, Nauka.

Lehninger, A.L. (1965). *Bioenergetics*. Benjamin, W.A. (Ed.), New York.

Lehninger, A.L. (1985). *The principles of biochemistry*. V. 1-3. Moscow, Mir.

Life of plant. (1974). In 6 vol. A.P. Tahtadzchayn ed. Moscow, Prosveshenie.

Margulis, L. (1970). *Origin of eukariotic cells*. Yale University Press, New Haven.

Mazia, D. (1961). *Mitosis and the physiology of cell division*. Brachet, J. and Mirsky, A. (Eds.), Academic Press, New York, London.

Merkis, A.I. (1990). *The force of gravity in the processes of plant growth*. Moscow, Nauka.

Rees, A.R. and Sternberg M.J.E. (2002). *Introduction to molecular biology: From cells to atoms*. Moscow, Mir.

Rubin, A.B. (1999). *Biophysics*. V. 1-2. Nauka, Moscow.

Shlegel, G. (1987). *General Microbiology*. Moscow, Mir.

Singh, H. (1978). *Embryology of Gymnosperms. Encyclopedia of plant anatomy* X (2). Berlin, Stuttgart, Gebruder Borntraeger.

Swanson, C.P. and Webster, P.L. (1977). *The cell*, 4th edn. Prentice-Hall Inc., Englewood, Cliffs, Jersey.

The cell cycle. Problems of regulation. (1973). O.I.Epifanova Ed., Nauka, Moscow.

Tokin, B.P. (1987). *General embryology*. 4th ed., Moscow, Visshay shkola.

Volkenshtein, M.V. (1983). *Theoretical biophysics*. Nauka, Moscow.

Wolf, S.L. (1993). *Molecular and cellular biology*. Wadworth Publishing Comp., Belmont, California.

Zhimulev, I.F. (2003). *General and molecular genetics*. Novosibirsk, Sibirskoe universitetskoe izdatelstvo.

Chapter V

Physics Place in Some Fundamental Processes of Biology: Hypothesis and Physical Models

Abstract

In this chapter the four physical models of different biological processes are described: the separation of organic order on a plants and an animals; the evolution of sexual gametes; the differentiation of gametes by sex, the optical activity of biological molecules and the role of the laws and principles of classical and quantum physics in all these processes. We have tried to see and understand the unknown physics in modern biology; the physical principle of the organization of plasmatic membranes, mechanisms of meiosis, etc.; the consequences of the dependence of biological processes on the main laws and principles of physics, namely, their influence on the evolutionary development, biological diversity and external appearance of living matter.

For progressive evolution, reclamation of dry land and searches of new sources of nutrients it was necessary to increase the operative function of the organisms. It is difficult to say, what a world of nature would be if its development did not submit to the laws of conservation of total momentum. The distribution of its action on the organic world has divided it on the two independent evolutionary branches of organisms - the animals and the plants -with different life strategies, and the sexual process, accordingly, on the two types of sex cells – the male and female gametes. We made a supposition about the quantum nature of sex and the sexual distinctions of male and female gametes. It is shown, during what moment of male meiosis and activating female meiosis there occurs a differentiation of gametes by the sex depending on the spin direction of the DNA. The physical model of the optical isomery of biomolecules and living matter as a whole is discussed.

1. The Low of Linear Momentum Conservation and Evolution of Biological Systems

It is known that matter without a motion not exist in nature. Life is one of the forms of matter motion, but it is qualitatively higher in comparison with its physical and chemical forms. At present all forms of matter motion are divided into three levels: the inorganic, biological, and social systems. Between levels there is a hierarchical subordination and a mutual influence. The change of material objects (atoms, molecules, cells, etc.) on each of them occurs as a regular and directed process, and in accordance with the fundamental laws of physics (the law of the corpuscular-wave duality of matter, the laws of conservation of the total energy, momentum, charge, etc.). The movement embraces all forms of the development of living matter, including evolutionary development.

Organic evolution, is one of the main forms of living matter development, and is characterized by a specific direction and irreversibility. The common ancestor of all animals and plants are the organisms of a prokaryotic organization level. The evolution of prokaryotes is the indivisible process. It began to diverge into two large independent branches - the animal world and the plant world – as a result of the transition of living matter on the multicellular level and during a mastering of the land. We tried to link the nature of this divergence with the manifestation of the law linear momentum conservation (or impulse). In the literature the changes of biological systems, which are caused by the actions of the law of linear momentum conservation, are not considered.

In classical mechanics, the linear momentum (**P**) of a material object is called the product of particle mass on its velocity. It is a vector quantity. This law for a large number of interacting particles is expressed as:

$$\mathbf{P_1 + P_2 + \ldots + P_N = const}, \text{ or}$$
$$\mathbf{M1v1 + m_2v_2 + \ldots + m_Nv_N = const.}$$

All prokaryotes are unicellular organisms. These are the open equilibrium quantum-mechanical systems with a simple enough intracellular structure. Their biological harmonious oscillator presents the cell together with an oscillating DNA. This type of oscillator is characterized by low quality and a poor energy level. Prokaryotic organisms are in close contact with the environment, and nutrients can to enter into a cell during almost any cell cycle. The predominant type of metabolism is fermentation, which in comparison with respiration and photosynthesis is a less effective way of reception of free energy. That is why the object of an evolutionary cell change has provided for itself. After fundamental changes, it has given rise to the eukaryotic cell or the cell system with a higher level of organization.

Unicellular eukaryotes are characterized by a higher level of orderliness of the intracellular organization, behavior, the ability for adaptation and developmental potency. For these purposes a lot of energy is required. Therefore during evolution, the energetics of a cell has undergone a number of essential changes. First, the biological harmonic oscillator has moved into the nucleus, due to a DNA-wave beginning to master the higher energy levels and living matter has passed to the multicellular level of organization. Second, the effectiveness of biological oxidation reactions in connection with the localization of biochemical systems in specialized organelles (respiration into the mitochondria, photosynthesis into the chloroplasts) is increased. Third, the metaphase chromosome as one more internal energy sources has appeared. Therefore, the energetics of the eukaryotes begin to be provided with greater number and more powerful sources of energy: by the two internal sources (a biological harmonious oscillator and metaphase chromosomes) and by the two outer sources in plants (respiration and photosynthesis) or the one outer source in animals (respiration).

The carrying over of a biological harmonious oscillator into a volume of nucleus and mitosis origination became the largest evolutionary achievement during the transition from prokaryotes to eukaryotes. It has led to the occurrence a greater number of phyletic lines on the basis of the unicellular organisms of eukaryotic level organization. They include different species: mobile and without specialized organs of motion; with a rigid external envelope and without it; contain chlorophyll or it is absent; with a haploid or diploid life cycle; with or without meiosis; etc. There are species, which unite at the same time as the structures of animals and plants in different variations. Apparently, the order of unicellular eukaryotes at an early stage of evolution was united and the distinctions between the animal and plant cell began to define only. Therefore the questions of the systematic affinity of a species have always served and serve as a discussion subject.

All unicellular eukaryotes spend a lot of energy on the maintenance of their viability, because of the large ratio of cell volume to its surface. Their life becomes more dependent on the nutrients in the environment, in spite of the fact that the eukaryotic cell spends the energy more efficiently. Therefore, they are compelled to feed constantly and search for new nutrition sources.

The transition of living matter on the multicellular level, development and differentiation of life forms and organs assumes the further growth of energy expenditures, which are necessary, on the one hand, for the maintenance of constancy and orderliness of the internal environment, and, on the other hand, for the perfection of bonds with an external environment. The life of individuals began to depend on the energy resources of external environment more. The speciation process on the two groups of species began to disperse depending on the mode of life, a way of extraction of nutrients and the paths that they use in a metabolism. In one part of a species the operative function for the search of the new sources of nutrition is more developed. The relations in a system of "predator-prey" have started to arise and be improved. As a consequence the world of live and lifeless matter has appeared interconnected among each other by means of the great number of food chains. In another part of the species the conservative function, on the contrary, becomes intensified. They became attached to a substrate, having passed into a motionless way of life and began to use the substances of "micro-niche" with great efficiency. As a consequence, the whole organic world was divided on the two independent branches – the animals (Metazoa) and the plants (Metaphyta). This separation accompanied by the divergence of structures at all levels of organization of genetic material. Distinctions between animals and plants became deeper and more obvious after the exit of them on land.

The analysis of a course of evolution showed, that the development and improving of organization and a divergence of species has a specific physical meaning. Let's look at the origination of animals and plants as in the global process which submits to the law of conservation of a total momentum. As it was mentioned, the living matter during transition to a multicellular level of organization has risen before the necessity of strengthening the operative function for the constant search of nutrition sources and a survival of individuals in the environment. If we proceed from an understanding of unity of the world, the change of mobility is possible only under the condition of the separation of all life in two parts. In order for the conservation law of impulse to be satisfied during evolution, an increase of mobility in one part of the biological systems should be compensated by the loss of the given function of another.

As the total momentum of living matter remains unchanged, the internal forces during evolutionary development are balanced in such a manner that:

$$\mathbf{mv}_{animal} = \mathbf{mv}_{plant.}$$

The separation of living matter on the two branches could occur in different ways, if we proceed from the formula calculating the impulse (**P**). The first way is, when the operative function of one part of a species (**v**) at a relatively constant mass (**m**) is progressively increased, and at another part of a species it is decreased proportionally. The second way is when the increase in mass of one type of species is compensated by its reduction of other species by the same mobility. The third way is when the growth of the mobility of one species is connected with the reduction of their mass, and, accordingly, the mobility of another species is lost with the increase of their mass. The result of realization of these three ways is the occurrence on the Earth of mobile, heterotrophic species of animals and motionless, autotrophic species of plants. (It should be noted, that a number of processes impact the morphological manifestations of the law of impulse conservation: the relations between generative and vegetative sphere of species, an information program of the genome, etc. which for simplification of analysis in the given chapter are not considered).

The submission of a speciation process to the law of momentum conservation has completely changed the appearance of living nature and its multicellular life forms. Within the eukaryotes originated two types of the species, which organisms are characterized by a different organization and vital strategy. In one type of species – the kingdom of Animalia or Metazoa – the vital strategy is directed on the intensification of the operative function of mobility (**v**) with a relatively fixed body mass (**m**). The morphogenetic differentiation of their organs occurs during early stages of embryogenesis. Accordingly, the number of cell divisions is at an end as they develop and the sizes of their bodies are limited. In comparison with the unicellular ratio of the surface of its body to its volume (or weight), which is in contact with an environment, as with multicellular animals became less. It promoted the decrease of energy expenses in the time of exchange with an environment.

This mobility has given to animals the essential advantage for the active search of trophic resources in order to extract from them energy, substance and utilization into their metabolism. Thanks to this, they have created unique functional systems (nutrition, digestion, osmoregulation, thermoregulation, gas exchange, circulation of liquids, allocation and coordination), which support their viability throughout their life. Animals have a more complicated organization with four types of main structures (plants have the only three types). The warm-blooded mammals have created a system of thermoregulation, which provides still higher levels of control of the internal environment.

The main part of animal cells is constantly in a state of active viability. They are characterized by weak vacuolization and haven't a hard (cellulose) envelope. Due to mobility their energetics are provided by three sources of energy: the two internal (a biological harmonic oscillator and metaphase chromosomes) and the one external (a respiration).

In the second evolutionary branch – the kingdom of Metaphyta – the speciation process according to the law of momentum conservation went in an opposite direction. The vital strategy of plants is directed on the intensification of the conservative function of species, a transition to a motionless way of life and an attachment to a substratum. If the mobility in plants is absent, then a speed aspires to zero, or:

$\mathbf{V}_{plants} \rightarrow \mathbf{0}$

From the law of momentum conservation follows, that the mass (**m**) of plants should increase proportionally with thc decrease of speed. Let's look, whether so it?

It is known, that a morphogenesis in plants proceeds during the entire life cycle. All higher order plants have a non-determined meristem – an apical meristem of shoot and root, the cells of which are divided by means of asymmetric mitosis. Each cell division of the given type finishes by a renewal of the genome of the maternal cell and in one of the daughter cell only (Chapter 3, see). This means, that a meristematic tissue can theoretically divide an uncountable number of times. Therefore a potentially possible number of cells, which meristematic cell can give birth during asymmetric mitosis, are planned out to infinity. Theoretically, their total weight will be immense as follows:

$\mathbf{m}_{plants} \rightarrow \infty.$

Actually, there are a number of the different internal and external reasons, which limit a growth and duration of plant life by the specific limits.

When plants grow in the same place for many years, it becomes much more difficult to obtain nutrients from the environment, than to extract them by the active manner, moving constantly on a vast territory of natural habitat. Therefore the plants absorb the nutrients continuously and superficially. In comparison with unicellular organisms the ratio between surface and volume, which being in contact with an environment, has increased many times. A body (a crone, root system) is strongly dismembered and the surface of their contact with an environment reaches considerable value. A rigid spatial orientation during cell divisions for the plants is characteristic. The morphogenesis or the path of transformation of the meristem in the differentiated, specialized and dead cells for each tissue is genetically defined and shorter. Each year, the differentiated cells in large quantities die, either turning in on the conducting tissues and supporting tissues, which is characterized by great durability, or they are removed during the falling away of leaves, fruits and flowers. A source of new cells is the thin layer of embryonic tissue (apical, cambial and root meristem), whose cells throughout life retain the ability to divide.

A vital strategy is reflected on the internal organization and functioning of the cell, because of the motionless way of life of plants. A plant cell in comparison with an animal cell is less mobile. During ontogenesis a somatic cells remain in one place and cells of every type of tissue have a strict spacing. They are covered outside by a thick and rigid cell wall. The differentiated cells are strongly vacuolated. A cytoplasm and the organelles occupy only a thin parietal layer.

An energy deficit is compensated due to the fact, that the plants have two outer sources of energy. The photosynthesis is the additional source, which is seen only in plant cells. The essence of photosynthesis is due to the ability of a cell to catch the radiant energy of sunlight and to transform it into the energy of ATP chemical bonds. For this purpose in a cytoplasm there are special pigments (chlorophyll), a complex system of photosynthetic phosphorylation and specialized organelles - chloroplasts. The analyzed morphological characteristics testify to the fact that the plants experience a permanent shortage of energy during long development in the same place in their environment. Possibly, therefore, plants do not have the system of thermoregulation and a considerable part of the cells in an adult condition are dead.

2. The Law of Linear Momentum Conservation and Evolutionary Development of Sexual Gametes

The law of linear momentum conservation during the evolution of sexual gametes is shown more evidently. However if the systematic groups of Metazoa and Metaphyta during evolutionary development have completely dispersed, the sexual gametes, morphologically changing, remain the interacting particles in fertilization. On the one hand, the gametogenesis for the study of the physical nature of the given phenomenon is more difficult, and on the other, it is more obvious due to the fact that the sexual gamete is a strongly modified cell, but not a multicellular organism.

The gametic dimorphism varies widely – from the absence of dimorphism (isogamy) up to its evident manifestations (oogamy). The type of sexual process, when fertilization represents the fusion of two morphologically identical gametes, is called isogamy. The isogametes are the active mobile or amoeboid sexual cell, and may have a flafellum. The isogamy was found in algae, lower fungi and many species of protozoa.

When there is isogamy, the gametes are identical in form, size and mobility. During copulation (fusion) a male gamete is distinguished from a female gamete by the inner organization and not by behavior. We have all bases to believe, that isogametes differ by their physical nature. Therefore in this description we will designate them as male (♂) gamete and female (♀) gamete as well. The hypothesis about the physical nature of sexual distinctions will be explained later in this chapter. Since the form, sizes and mobility of isogametes are identical, then:

$$\mathbf{m}_{\text{♂}} = \mathbf{m}_{\text{♀}}\text{и}\mathbf{v}_{\text{♂}} = \mathbf{v}_{\text{♀}}.$$

Accordingly, the momentum of isogametes has the same quantity:

$$\mathbf{m}_{\text{♂}}\mathbf{v}_{\text{♂}} = \mathbf{m}_{\text{♀}}\mathbf{v}_{\text{♀}}.$$

The total momentum of the two gametes (male and female), is recorded as:

$$\mathbf{m}_{\text{♂}}\mathbf{v}_{\text{♂}} + \mathbf{m}_{\text{♀}}\mathbf{v}_{\text{♀}}$$

- is a constant quantity. I.e., the internal nature of sexual process are organized in such a manner that the total momentum of the two types of gametes during a time cannot change. We know from physics, that any bodies, which possess an equal velocity and mass, are stopped and stick together at impact. The same behavior of gametes occurs before and during fertilization. The male and female gametes of Chlamydomonas, mutually attracted and approaching, are moved toward each other at a specific velocity. They stop after the collision and stick together by means of the "nouses". Then the process of their syngamy begins. The envelopes of the two gametes are dissolved, the contents of a male gamete enters into a female gamete, but is already under the influence of other forces, the nature of which will be described below.

An evolution of the sexual process occurs simultaneously in the two kingdoms of Metazoa and Metaphyta from an isogamy to the anisogamy and then from an anisogamy to the oogamy. The expediency of such changes is caused by the necessity of the development of the operative function or the ability for the independent locomotion of sexual gametes. The origination of gametes, which are characterized by energetic motion, essentially decreases the time for a partner search, increases the probability of a meeting of gametes and the efficiency of the reproductive process. If the given process submits to the law of conservation of momentum, then the changes of one type of gamete should occur with the exact coordination of the changes of other gamete type. According to Newton's II law the speed of an impulse change of the male gametes (♂) during this time should be equal and oppositely directed to the speed of an impulse change of the female gametes (♀), or:

$$\frac{dPmale}{dt} = -\frac{dPfemale}{dt}.$$

Since at present only the internal forces of the total momentum of gametes cannot change, then a number of consequences from here must follow. First, the evolution of a sexual process must obey the law of linear momentum conservation. Second, the isogamy in the course of progressive evolution should give rise to the two types of gametes – male and female, which differ by their mobility. Third, during speciation the speed of the change of mass and mobility in both types of gametes are strictly coordinated in space and time. I.e., if the impulses of the two types of gametes add, then the speed of change of the sum under the influence of only internal forces is equal to zero. Therefore when they differ by size and by the form of their eggs (or ovicells) and sperm cells before fertilization approach each other, they at the moment of collision are stopped and stick together.

The first step on the path of the operative function change of sexual gametes is the emergence of anisogamy. At the anisogamy they differ in size, form and mobility takes place already as well. The microgametes have the flagella and they are the most mobile and smallest on the size scale of generative cells. On the contrary, the macrogametes are characterized by their larger sizes. They may be of the mobile (many algae) and motionless (macrogametes of many protists) forms. Since in the system of two interacting sexual cells the sum of their momentum is conserved, then:

$$\mathbf{m}_{\male}\mathbf{v}_{\male\,\mathbf{microgametes}} + \mathbf{m}_{\female}\mathbf{v}_{\female\,\mathbf{macrogametes}} = \mathbf{const}.$$

The dependence of the evolution of a sexual process on the law of the conservation of total linear momentum is well traced both within the one type of gametes, and between different types. In the male gametes (microgametes) the speed of the increase of mobility occurs proportionally to a decrease of their mass. In the female gametes (macrogametes) the given process goes at the same speed, but in an opposite direction – the reduction of mobility is compensated for at the expense of the increase of gamete mass. This means:

$$\mathbf{m}_{\male}\mathbf{v}_{\male\,\mathbf{microgametes}} = \mathbf{m}_{\female}\mathbf{v}_{\female\,\mathbf{macrogametes}}.$$

During passage to the oogamy stage the divergence of male and female gametes by their mobility, form, mass and metabolism has decreased. The fertilization of the given type of sexual process takes a large part, motionless ($V_{\female} \rightarrow 0$) female gametes – eggs (ovicells), and very small, actively mobile male gametes – spermatozoids (or sperms in high order seed plants, which are delivered to the ovicell by means of pollen tube).

The mature eggs and sperm cells are very changed and high-specialized cells. If they fall away from the reproductive system, then, as a rule, soon perish. An egg represents the motionless cell, which has grown to huge sizes. It usually has a rounded or oval form, and less often a lengthened form. In comparison with a somatic cell it is characterized by a haploid set of chromosomes and a powerfully altered nuclear-cytoplasmic ratio. The metabolism of a female gamete, the balance between mRNA, tRNA and rRNA to the advantage of rRNA is broken; the quantity of DNA on a nucleus in some species is increased. The cytoplasm is filled completely by the nutrients – the yolk inclusions which are spent by an embryo during embryogenesis. A mature egg is a metabolically inactive structure. The RNA, DNA, proteins, lipids and polysaccharides are in a condition, which is inaccessible to recycling. The assimilation dissimilation processes are brought to a minimum level.

During evolution the male gametes are differentiated in the opposite direction. The sperm cells are very small in size and the mobile cells have a haploid set of chromosomes. They are characterized, as a rule, by a very elongated and pointed shape and often have the flagella. A flafellum is the long, threadlike organelle. The sperm cells may be without flagellate or have, one-, two-, or many-flagellate types. The elongated shape and flagella promote a more active motion of sperm cells and increase the speed of locomotion in a liquid medium, which to them is a vital environment before fertilization (except the higher order plants). The nuclear-cytoplasmic ratio of male gametes is also changed, but in the opposite direction. The main part of the cell volume occupies the nucleus, with a small amount of cytoplasm and nutrients.

Up until now in nature, it has not revealed any case of a violation of the law of conservation of momentum, between the speed of male gametes and the mass the same ratio remains, as in the practically motionless ovicell, which has grown to a huge size. In this case the ovicell during fertilization will seem to be standing in one place, and the sperm cells bombard it with great velocity. Nevertheless the same result takes place at the collision of gametes, as during isogamy and anisogamy.

It is necessary to underline, that in the historical plan of the transition from isogamy to oogamy, as well as the processes of the separation of living matter in animals and plants, have occurred rather quickly (in plants within the division of *Chlorophyta*, in animals within the division of *Protozoa*). The further development of multicellular organisms occurred and continues to occur in accordance with the law of conservation of momentum on the basis of only one type of sexual process - oogamy and in parallel in the two different kingdoms - Metazoa and Metaphyta, provided that:

$$\mathbf{P}_{\male\,\mathbf{gamete}} + \mathbf{P}_{\female\,\mathbf{gamete}} = \mathbf{const} \text{ and } \mathbf{P}_{\mathbf{enimals}} + \mathbf{P}_{\mathbf{plants}} = \mathbf{const}.$$

A male and female gamete differs not only by mobility, size and external appearance, but by the internal state of genetic material also. As a rule, this type of differentiation is described as a definite sexual balance. A hypothesis of the possible physical nature of sexual distinctions will be discussed below.

3. Hypothesis of the Quantum Nature of the Sexual Distinction of Male and Female Gametes

At the beginning of this chapter, we have tried to understand the logic of the development of evolutionary processes from a position of the laws of classical physics. For an explanation of the nature of a sexual distinction between male and female gametes we have long searched for such theory, which coincides with the experiment. Eventually we paid attention to the quantum mechanics. The proposed physical model describes the nature of the sexual distinctions of gametes as a purely quantum-mechanical phenomenon, which is determined by the inwardness of a molecule of DNA and can change under specific physical conditions.

The nature of sexual distinctions always was and will be among fundamental and the most enigmatic problems of biology. The structures, mechanisms and processes associated with sex are various and problematic. However, it is obvious, that the mechanism of the sexual differentiation of gametes is necessary to search somewhere inside the structure of the reproductive cycle or any one of its stages. In and of itself, the reproductive cycle of every species is a conservative formation. Absent of casual cells there is not a single unnecessary cell division. Each generative cell has a specific corpuscular-wave characteristic, spacial location within the reproductive complex, time origination, functional mission and destiny. During reproductive development the separation of gametes by sex occurs as well.

The notion of sex usually includes two different phenomena: a sexual process and a sexual differentiation. Both processes create a sexually-linked variety. However, a sexual process and a sexual differentiation are the phenomena that are inherently opposite. The sexual process creates a variety on the basis of haploid male and female gametes. The sexual differentiation is characterized not by less diversity but on the basis of diploid bisexual and dioecious organisms.

From the range of problems related with sex, we have selected as an example only one question – the sexual differentiation of gametes during usual (or male) and activating (or female) meiosis. At present the biology of these types of meiosis is studied in detail enough, that is what allows us to understand its essence from a position of quantum physics. At the time of the writing of this chapter the data from the research of several sciences has been used: development biology (Poddubnaya-Arnoldi, 1976; Singh, 1978; Kozubov et al., 1982; Embryology of Angiosperms, 1984; Tokin, 1987; Belousov, 1980, 2005; Zhuzhzhalova et al., 2007; Gilbert, 1910; et. al.), systematics (Komarnitskii et al., 1975; Life of plant, 1978;Vasiliev et al., 1988; Reyvn et al., 1990; et. al.), cytology of meiosis (Darlington,1965; Petrov, 1973; Cytology and ..., 1975; Sybenga, 1975; Swanson and Webster, 1990; Bogdanov, 1991; Alberts et al., 2002; Chentsov, 2004; et. al.) and the Feynman lectures on physics (Feynman, et al., 1967). As known, the quantum mechanics are characterized by a very difficult mathematics. Therefore our task was to describe the conception so that a difficult form of physics was accessible to the majority of biologists, and the biological and physical essence of the analyzed processes has not been lost.

A molecule of DNA and a nucleotide, as its monomeric links, can be considered from two positions - a part and a whole. DNA, as the category of "whole", represents the chemical compound – a molecule of deoxyribonucleic acid. As the category of "part" this high-molecular polymer consists of two spirally twisted polynucleotide chains and a great number of repeating monomeric units – nucleotides. Accordingly, a nucleotide is the independent deoxyribonucleotide link. Into its composition enter (as a complex micro-object) the nitrogenous base, monosaccharide and phosphate groups. Therefore, the DNA and its nucleotides are simultaneously both indivisible and complex micro-objects.

The notion of spin completely pertains to quantum mechanics. A spin in the enclosure to the complex and composite micro-objects is accepted to define how the intrinsic angular momentum of a quantity of motion is in a condition of rest, which is not connected with the micro-object motion as a whole. Proceeding from the universal principle of the organization of matter, we have made the supposition, that a nucleotide and molecule of DNA as complex and indivisible micro-objects possess the certain spin (as an electron, a proton, etc.).

Let us assume further, that a DNA and nucleotide are the quantum-mechanical micro-objects, which behave as the Fermi particles (we proceed from the fact that, during reproductive development only the two types of particles are formed – the male and female gametes). If this is so, the two complementary nucleotides, laying in two polypeptide chains against each other and the bases of which are connected by the hydrogen bonds, cannot have the identical direction of spin. This follows from the Pauli exclusionary principles that two identical Fermi particles cannot get into one and the same state, because in this case the interference of their amplitudes will be with a negative sign. From here this important consequence follows – the spins of complementary nucleotides should be oppositely directed only.

The existence of a spin-spin interaction has been proved in the middle of the 20th century. In the DNA, due to a spin-spin interaction between the nucleotides, the ordered orientation of intrinsic angular momentum of a quantity of motion extends onto other nucleotides. This leads to an establishment of a stable and uniform orientation of spins along the entire length of a polynucleotide thread of DNA. Since the spins of two threads of DNA are directed to the opposite sides, between them arise the forces of mutual attraction. Hence it follows, that the stability and conservatism of a macromolecule structure is achieved and supported, perhaps, by means of two forces – the specific hydrogen bonds between the complementary bases, and a spin-spin interaction between two threads of DNA.

Accordingly, during replication there should not be a violation of orderliness of the spin system inside of doubled chromosomes. If we proceed from the Pauli exclusionary principle and the corpuscular-wave mechanism of a DNA biosynthesis (see chapter 2) it can occur as follows (Figure 31, 32). The negatively charged enzymes of the DNA-polymerase are connected to the different chromosomes-waves. During mutual repulsion the two threads of DNA in the places of association with enzymes move away from each other and as a result a specific number of replication folks appear on the DNA. The hydrogen bonds between the threads locally break, and the forces of a spin-spin attraction between the two DNA threads in this zone become weaker. A DNA-polymerase, moving together with the wave, synthesizes the daughter molecule of DNA on the complementary basis. The two nucleotides can draw nearer to each other and between them a chemical bond is formed if their spins are directed towards each other. Therefore one thread of parental DNA synthesizes the daughter thread, nucleotides of which are directed with the spin upwards, and its second thread, accordingly, with the spin downwards.

Thus, each chromosome after a replication consists of four Fermi particles of two types – with the spin upwards (↑) and with the spin downwards (↓). According to the Pauli exclusionary principle these four Fermi particles up to the beginning of mitosis should occupy a specific position inside of the doubled chromosomes and formed the two pairs of macromolecules, as is shown in figure 32. One thread in each pair belongs to the initial parental DNA, and the second thread is the newly synthesized daughter DNA. The spins of two the DNA molecules within every pair are directed towards each other. The DNA molecules in each pair are connected to each other by means of the hydrogen bonds and the forces of a spin-spin interaction, while between the pairs are formed only the hydrogen bonds. It leads to the fact, that the DNA-particles inside a pair are more tightly connected, than between pairs. I.e., it is one more reason which explains why the splitting of chromosomes during mitosis should be carried out by the hydrogen bonds between pairs, or by the semi conservative way of reproduction.

Meiosis

A sexual reproduction in the corpuscular biology is only associated with the meiosis. Meiosis is a special type of cell division, which leads to the formation the four haploid cells, differentiated on the male and female gametes, from one diploid cell without the sex differences. Consequently, the nature of sexual differences should be found in the mechanism of meiosis. For this we must to go back to the corpuscular-wave model of meiosis, which was described in Chapter 3. Let us consider its mechanism, taking into account, that each DNA thread is a complex and indivisible micro-object, which behaves as a Fermi particle. The analysis will begin from the stage of leptotene, when the double set of chromosomes made a turn and changed the spatial disposition inside a nucleus.

It should be noted, that at the description of meiosis in Chapter 3, for a simplification of the model we have deliberately missed the two facts, which are connected with the definition of the sex of gametes. First, that during leptotene the turning of the chromosome set and the change of their orientation takes place concerning both a direction of gravity and the magnetic field lines of the Earth. Second, those centrioles are mobile structures, which, like other organelles, actively circulate inside a nucleus until the biological harmonious oscillator operates. The oscillator is demolished during mitosis prophase or cell death. Now let us see, what consequence follows from the fact, that centrioles are the moving charged particles and what occurs during their motion in a cell, which is the beginning of a meiotic division.

According to the laws of physics, the two positively charged centrioles, moving in a circle into a nucleus, generate an electric and a magnetic field. This internal electromagnetic field interacts with the external stationary magnetic field of the Earth. The interaction is then amplified or weakens, depending on the direction of vectors of the intensity of external and internal fields. As a result, the intensity of the total field is constantly changing.

We initially assumed, as you will remember, that the nucleotides are the Fermi particles. During leptotene-pachytene (G_2 period) the nucleotides of doubled DNA with specific periodicity pass through a non-uniform magnetic field, which is created by the two positively charged centrioles and the magnetic field of the Earth. The magnetic field interacts with each nucleotide and they under its action are filtrated into two separate beams in depend of the direction of the spin. The phenomenon of the splitting of atoms on two beams in a non-uniform magnetic field, depending on the spin direction was proved in the well-known experiments of Stern - Gerlach. The mechanism of filtration by the spin in the case of four nucleotides is shown schematically in figure 31.

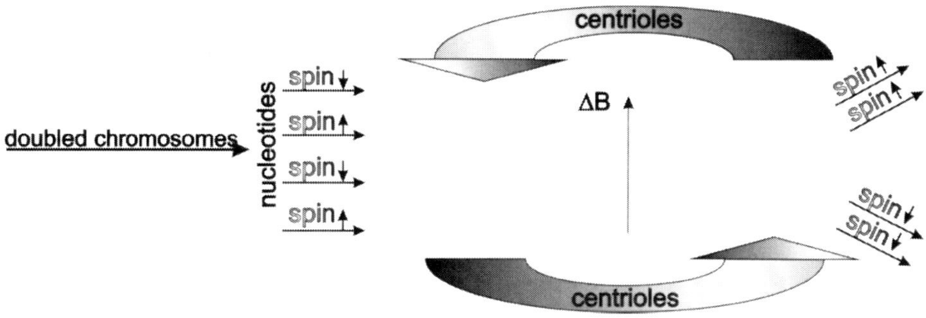

Figure 31. The scheme of nucleotides filtration, as the Fermi particles, into a non-uniform magnetic field and them splitting on two separate beams in depend of spin direction.

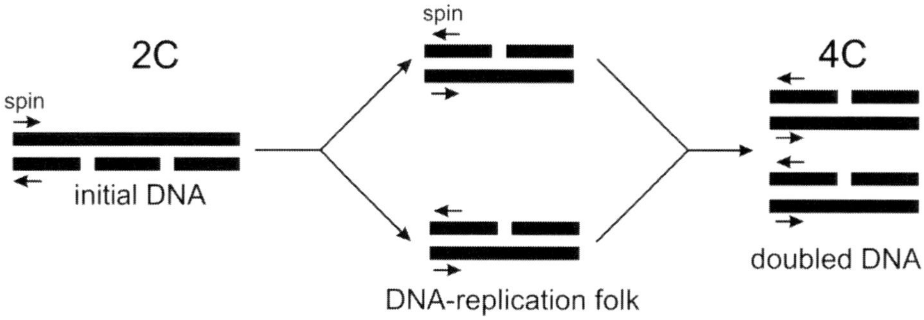

Figure 32. The probable mechanism of replication, if a DNA is the Fermi particle.

The filtration of the nucleotides on the two beams leads to the origination of an exceptional structural formation – bivalent-particles and bivalent-waves. The bivalent-particles are the pair connections of doubled homologous chromosome-particles. The bivalent-wave is the resultant wave arising during the interference of waves in the areas of chromosome coiling. The formation of bivalents begins with a turn of a chromosome set into nucleus and changes of spatial orientation of chromosomes in leptotene (see Chapter III). After that the homologous chromosomes are disposed territorially in close proximity of each other. During zygotene the homologous chromosomes are drawn to each other, giving rise to the bivalents. The pachytene is the period of maximum conjugation of chromosomes, when their dual nature becomes invisible (morphological effect of the interference of homologous chromosomes-waves).

Now the processes of bivalent formation from a position of quantum mechanics will be analyzed. A DNA coiling begins after the process of replication and continues up to the pachytene end (Figure 33). During meiosis the rotation of isomers and a conformational change of the macromolecule are connected with the chromomeres, the number and location of which on a body of chromosomes is characterized by its constancy. In a result, after each turning the length of a chromomere becomes more on one link, the DNA molecule has decreased a little, its potential energy is increased and the complicated new isomeric link is formed. The next events occur simultaneously, when the four nucleotides enter into the zone of an electric and non-uniform magnetic field, is the filtration of nucleotides. Therefore every new isomer link is formed on the base of filterable nucleotides. When the process of chromosome spiralization is ordered, the attaching of the next fragment does not lead to the violation of the mutual orientation of the spins inside each chromatid and chromomere site or the chromosomes as a whole.

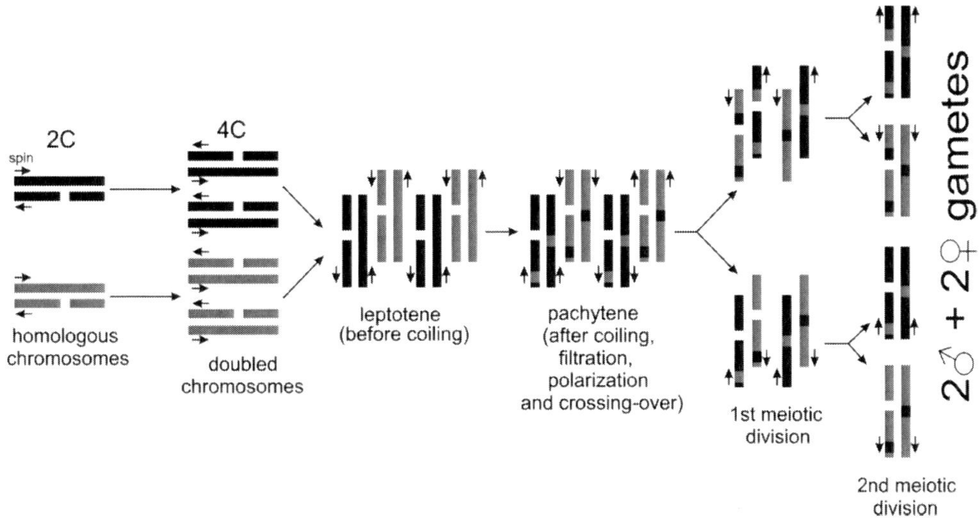

Figure 33. The probable mechanism of formation of male and female gametes during meiosis.

Thus, during leptotene-pachytene a DNA thread is gradually shortened, thickened and polarized. By the end of pachytene (after filtration of the nucleotides and spiralization by means of internal coiling) it consists of the two piles of complex isomers: one of them contains the two identical polarized threads of DNA (future chromatids) with a spin up, and another, accordingly, the two threads with a spin down (Figure 33). The identical polarized threads of the DNA are kept together by means of the forces of the electrical attraction between the nucleotides. Between the homologous chromosomes arise the spin-spin interactions, which also stabilize the internal structure of the bivalents. The macromolecule during filtration acquires the new quantum-mechanical properties. Morphological confirmation of these properties is the attraction of the homologous chromosomes to each other and the origination of bivalents during meiosis.

Such a mechanism of bivalent creates a unique situation from the position physics, which is contrary to the Pauli exclusionary principles. I.e., in bivalents are located in close proximity to each other the two chromatids, or two fermions with the same spin direction. To understand the given situation, we will imagine that in a bivalent one chromosome is an "internal particle" and the homologous chromosome relative to it will be an "external particle". In this case, the relations between the "internal" and "external" nucleotides of homologous chromosomes should be the following: the complementary pairs of nucleotides with the spin-up should be placed opposite of the adjacent pairs of homologous nucleotides with the spin-down. This is caused by the sister chromatid, aspiring to counterbalance the parallelism of the spin according to the exclusion principle, and uses for this purpose the intermediate environment – the sister chromatid of its own homolog. In other words, the tendency of the spins of the two DNA threads being parallel is compensated for by means of two homologous DNA threads with the opposite spin (Figure 33). The spin-spin interactions, arising between chromatids, attract the homologous chromosomes to each other, giving rise to the new structural formation - a bivalent. We conclude that the origination of bivalents during meiosis is due to the manifestation of the action of the Pauli exclusionary principle.

Another confirmation of the quantum-mechanical nature of bivalents is the behavior of the chromosomes during diplotene. If during pachytene the complete conjugation of homologous chromosomes is observed, then during diplotene-diakinesis the bivalents with the four thread structure appear, and the succession of their destruction is observed.

In a corpuscular-wave model (see Chapter 3) a bivalent disintegration initiates the disappearance of the DNA-wave and the biofield. The process of a continuous spiralization of chromosomes during leptotene-pachytene leads to them progressively shortening (internal coiling). At that moment, when the length of the chromosomes becomes less than the diameter of the nucleus, the DNA-wave and biofield disappear and two positively charged centrioles, mutually push off and scatter in different directions. At this time, first, there is a splitting of chromosomes on the chromatids up to the centromeres. Second, since the wave function on the DNA disappears, then a bivalent as the particle, does not obey the laws of quantum mechanics. Therefore the forces of the spin-spin interaction between the homologs cease to act. The homologous chromosomes continue to be kept together by means of orientational interaction between the piles of complex isomers, which have a polar structure. The divergence of homologs occurs only in a zone of centromeres, because they, as the two negatively charged structures repel each other.

During the first diplotene-metaphase a coiling of chromosomes occurs and each chromatid on the bases of the complex isomers is twisted independently (external coiling). Chromosomes decrease in size, become thicker, and their shoulders gradually lose a dipolar structure. I.e., the last force, that was holding the homologous chromatids within a bivalent, disappears, and they begin to move away from each other. At this time only the chiasm, or the sites of chromosomes without coiling, hold the homologs together. During diplotene the chromatids of homologous chromosomes are connected among themselves by a specific number of chiasm in the centromeric region and along the entire length of them. However, in the course of coiling during diakinesis the chiasm are displaced to the terminal ends of the chromosomes, the number of them decreases and the length of the bivalents becomes shorter. During metaphase a set of bivalents are moved in an equatorial plane of the spindle. After that they undergo two divisions and give rise to the four haploid sexual cells – two male gametes and two female gametes, which differ by size, morphology and spin direction.

Hence it follows that a pool of gametes of any species after reconstruction of the nucleus envelope and the wave function on a DNA (if the meyospore does not pass into a state of rest) will be present in the polarized quantum-mechanical systems of two types –one half with spin directed upwards, and the other half with spin directed downwards. Thus, during meiosis the sexless diploid maternal cell gives rise to four haploid cells with an identical DNA-wave, various DNA-particles (changed during crossing-over) and different sexual signs depending on the direction of DNA spin. The corpuscular-wave model of the quantum-mechanical nature of gametes and meiosis explains why the meiotic division is ended by a creation of a pool of sexual cells, made up of 50 % male gametes and 50 % female gametes.

We can assume that if a sexual gamete touches the spin objects, then they should interact with each other through a separating medium (water, air, a diploid tissue of nucellus, etc.). Examining the fertilization process from these positions, it becomes clear, when and with participation of what forces the syngamy of gametes is achieved and in which are not.

Now we will consider, how the hypothesis of the quantum-mechanical nature of sexual distinctions is confirmed by the experimental data of the embryology of animals and plants. The analysis will begin with the most known fact in embryology, that during fertilization the unisexual combinations of gamete crossing does not occur (♂ + ♂ and ♀ + ♀). Proceeding from the quantum-mechanical nature of male and female gametes, the answer to the given question is obvious. Indeed, two male and two female gametes, as the two fermions, which have the same impulse, and an identical direction of spin cannot approach each other, because in this case the total of their amplitude becomes equal to zero. By the Pauli exclusionary principle, during sexual reproduction mutually only the gametes of opposite sex (♂ + ♀) with spins pointing towards each other should be mutually attracted and participate in fertilization For a better understanding of the copulation mechanism it is necessary to take into consideration one important factor without which it is impossible to understand the nature of the behavior of the sexual gametes during fertilization. As was already stated, each cell as the quantum-mechanical system possesses its own electromagnetic field or biofield. During telophase it participates in the formation of the envelope of the nucleus and the cell, pushes the daughter cells away from each other and prevents nucleic fusion after mitosis. Because the forces of a spin-spin attraction act over a shorter distance, than the forces of electromagnetic repulsion, the biofield serves as a serious barrier for a syngamy of male and female gametes. It seems obvious, that fertilization is possible only in that case, when the forces of the spin-spin attraction between gametes are more than the forces of the repulsion of their biofields.

During a study of the sexual process in plants we have revealed the 3 types of the copulation mechanism of sexual gametes. Let us consider each of these in more detail.

1. Syngamy by the Conjugation Type

The given mechanism is the most primitive and converges in *Mougeotia* and in some species of *Zygnema* (class Conjugatophyceae). In a favorable environment the numerous threads of algae can develop near each other for long periods. The attraction between the cells and threads of different sexual valences is absent. In adverse conditions (the high and low temperature and the high intracellular pressure) the situation is changed. In the autumn when the water in the basins cools down, the tensity of the biofield around the cells of both sexual valences (i.e., with a spin-up and with a spin-down) begins to fall. The threads are attracted and positioned parallel to each other, when the biofields of opposite cells are diminished in this way the forces of the spin-spin interaction start to exceed the forces of the electromagnetic repulsion. A cell fusion is realized in the copulation channel, or in the equidistant zone from the initial cells, where the forces of electromagnetic repulsion are at a minimum.

2. Syngamy by the Isogamy Type

This type of fertilization mechanism is encountered in many systematic groups of the lower order plants (Ulotrichophyceae, in some species of classes Conjugatophyceae, Volvocophyceae, etc.). The transference of the fertilization process inside the cell is the first step in the evolution of a sexual process in the given direction. It is speculated, that a generative cell during fertilization attracts to itself the other generative cell, that is a female gamete. The female gamete possesses a higher magnetic energy. Accordingly, the generative cell, which it is attracted to, is a male gamete with lower magnetic energy.

We can assume that the origination of a given mechanism is caused by the differentiation of the cells by a spin, to be more exact by the alteration of the energetics of the male and female gametes in depend of the spin state. The living matter is very sensitive to a spin direction. Even a small shift of in an energy state of sexual gametes become stronger during the growth of the biological energy of some species and is affected by the organization of the reproductive cycle and the morphology of the generative cells. Therefore the process of copulation begins to occur in that gamete, a volume of which is increased progressively and the evolutionary development of the sexual process went from an isogamy to the anisogamy.

Continuing on from the nature of the interaction of sexual gametes, the fertilization mechanism in the isogamy type can be the following. In a favorable environment the forces of repulsion on the electromagnetic fields prevents the approaching of gametes in the distance, where the forces of the spin-spin attraction operate. As a result of the temperature regime change an attraction becomes stronger than the repulsion. A female gamete, as the more energetically strong cell in the spin-spin interaction of gametes, begins to carry over the contents of a male gamete into its own volume. Therefore during isogamy, anisogamy and conjugation in some species the protoplast of one cell completely passes into the other cell, where two nuclei copulate, giving rise to a zygote.

3. Syngamy by the Oogamy Type

This type of fertilization mechanism is more progressive and widespread among the lower order plants and occurs in all species of higher order plants. The oogamy is characterized by: 1) a formation of one (or a few) large motionless ovicell and the numerous small, mobile spermatozoids (sperms); 2) an appearance of special reproductive organs – an oogonium and an antheridium where the female and male gametes are developed; 3) a bifurcation of the reproductive cycle in the female generative line and the male generative line; 4) a female energy line originates from the higher energy levels, and a male energy line – from the lower energy levels. During fertilization the large motionless female gamete passes into a rest state. The biofield intensity gradually decreases and shrinks to a nucleus. Because during a rest state the electromagnetic protection is absent, the ovicell should be in the condition of maximum sexual receptivity. The small and mobile male gametes by means of the forces of a spin-spin interaction are attracted to a female gamete. The more energetically strong spermatozoid easily penetrates the external envelope, quickly overcomes a layer of cytoplasm, enters into the nucleus and leads the ovicell out of a rest state. The arising biofield instantly restores the electromagnetic protection of a fertilized ovicell and prevents, thus, the penetration into it by other male gametes (a spermatozoids, or sperms on higher order plants).

The diatoms are a good model for an understanding of the mechanism of fertilization in diploid organisms. As it is known, in *Centrophyceae* the typical sexual process is absent. An autospore is completely formed in the place of a maternal cell. Its origination is preceded by two divisions of the nucleus by means of the activating meiosis, in which takes place a "rejuvenation" of the genome. During reducing division the two haploid nuclei receive a portion of biological energy, and their DNA-waves pass to a higher level of energy. The other two haploid nuclei, on the contrary, lose this amount of energy and the oscillators in them "are discharged". During leptotene-pachytene meiosis a process of the filtration of nucleotides and the polarization of DNA by the spin direction occurs. As a result the formation of two nuclei with spin-up and two nuclei with spin-down takes place. Two haploid nuclei with the "discharged oscillators" and opposite spins die off. The "rejuvenated" nuclei under the conditions of high intracellular pressure, when the forces of the electromagnetic repulsion decrease and begin to interact with each other by means of the Pauli exclusionary principle. Because they have opposite directional spins, the forces of attraction are enough, in order for the nuclei to have fused and have given rise to the "rejuvenated" autospore.

The forces of a spin-spin interaction can estimate by cytological events during conjugation– the special type of the sexual reproduction in *Conjugatophyceae*. So, in *Spirogyra* during activating meiosis four haploid nuclei are formed, from which three nuclei are destroyed, and only one ("rejuvenated") begins a new cycle of ontogenesis. After a resting period the meiospores germinate and give rise to two kinds of organisms in depend of of spin direction – 50% with a spin-up and 50% with a spin-down. A vegetative generation of *Spirogyra* presents the multicellular threads with oppositely directed spins. In adverse conditions (a high and low temperature) it leads to a strong mutual spin-spin attraction, which is transferred through a layer of water. At the beginning the threads are arranged parallel to each other. Following this step is the occurrence of protrusions and a formation of the channel between the two cells. Then the contents of one cell (a male sexual cell) pass through a channel in the opposite cell (a female sexual cell), where the copulation of gametes takes place and causes the formation of a zygote. Naturally, the forces for the destruction of the cell wall and the creation of a channel between the cells require large amounts of energy. But much more energy is required so that the protoplasm of the male gamete can be "pumped" inside the female gamete and the contents of the two cells can compress into the volume of one cell.

In angiosperms between an embryo sac and a pollen grain the diploid tissue of a column is located. If we proceed from the experimental data on a spin-spin interaction between male gametes and female gametes, apparently, occurs through this stage as follows. During maturation and pollination the pollen aspires towards a rest state. Though after a hit on the stigma it comes out of the rest state, germinates and forms a pollen tube. The generative nucleus or sperm and nucleus of a vegetative cell pass through a pollen tube. Apparently, somewhere during same time the spin-spin interaction begins to act between the pollen tube and the embryo sac through a diploid tissue, which separates them, but only if the spins of given cell complex are located towards each other. That is why a pollen tube grows by its tip. The generative and vegetative cells are placed in this zone as well. In the beginning pollen tubes grow in a column (over the surface of the channel walls, in the channel, or between its cells). During growth they overcome a powerful resistance from the cells, however the force of spin-spin attraction is so strong, that the tube moves apart, often destroying nearby cells. The diploid tissue of the column carries out the selection of pollen tubes on the interspecific and the intraspecific incompatibility, and the forces of the spin-spin attraction makes a choice based on the degree of the polarization of the male and female gametophytes. As far as advancement of the pollen tube towards an embryo sac the forces of attraction between them will increase and the leading tube is allocated gradually. Then they pass into the ovary and continue the growth in the direction of an embryo sac. The basal part of the tube is periodically cut off by means of callose stoppers. All during this period the processes of incompatibility and the forces of spin-spin attraction participate in the selection of the leading tubes. The leading pollen tube approaches an embryo sac through the micropyle (or the chalaza), enters into one of synergids, passes through it and gets into a cavity of the gametophyte somewhere between the ovicell and the central cell. An ovicell from a condition of rest helps to create the alterations of physical conditions during the passage of a pollen tube through the envelope of an embryo sac and the function of the DNA-wave renews.

When the male gamete with a specific direction of spin and momentum P approaches a female gamete with the opposite spin, which is in the resting state, the attractive forces will be enough for the creation of the diploid set of chromosomes. For this the nucleus of a male gamete passes into the volume of the ovicell and a sperm crosses its cytoplasm and penetrates into the nucleus where their genetic materials are united together. The two haploid sets of chromosomes with opposite spins into a nucleus become closer to each other, only if their DNA-particle and DNA-wave have the identical corpuscular-wave characteristics. In this case a new individual and species-specific genome of a zygote (or after correction of DNA characteristics the other initial cell of a future sporophyte) arises with a double set of chromosome-particles and with a "recharged" set of chromosome-waves, which do not break up during a mitosis and, as the stable genetic structure, gives rise to a viable diploid plant.

It seems obvious, that in all species of the higher order plants and animals there are genetically bisexual organisms. When the female gamete with a specific direction of spin and momentum P approaches a male gamete with the same impulse P and an opposite spin, the attractive forces will be enough for the connection their haploid sets in the diploid set of chromosomes. For this the nucleus of a male gamete passes into the volume of the ovicell, and a sperm crosses its cytoplasm and penetrates into the nucleus where their genetic materials are united together. The two haploid sets of chromosomes with opposite spins into nucleus become closer to each other. Thus, a new (individual and species-specific) genome of a zygote arises with a double set of chromosome-particles and with "recharged" set of chromosome-waves, which do not break up during a mitosis and, as the stable genetic structure, give rise to a viable diploid plant.

Thus, the genomes of diploid organisms present a stable structure, because the spins of the homologous chromosomes look towards each other. It is experimentally proven that sexual dimorphism is determined immediately after fertilization and depends on the properties of the gametes participating in the fertilization. If we proceed from the basis of the quantum mechanical nature of sex, when during fertilization is possible only two combinations of gametes in depend of spin direction: 1) "♂ ↑ + ♀ ↓ or ♂ ↓ + ♀ ↑" for amonoecious bisexual species of plants and animals, and 2) "♂ ↑ + ♀ ↓» + «♂ ↓ + ♀ ↑" for a dioecious species of plants and animals. In the latter case, there 50% of the individuals are of the male sex, and 50% of the female sex. We can assume that the explanation of the nature of the sexual dimorphism and the mechanisms of its functioning are necessary to investigate the organization of these two systems. For this we turn to the data on the biology and evolution of gametes and cell systems associated with sexual differentiation.

Evolution of Sexual Gametes and the Sexual Process in Animals and Plants

As has already been stated, the mechanism of gamete differentiation by sex in our model has a quantum-mechanical nature. During meiosis in the local zone between two centrioles is created an inhomogeneous magnetic field, which during leptotene-pachytene separates the nucleotides into two polarized streams in depend of the spin direction. i.e., the mechanism of sex determination is a component part of meiosis. The sexual system originated simultaneously with meiosis and a sexual process and advanced in a long and independent way of evolutionary development. At the early stage the energy level of gametes was not yet related with their morphology and mobility. The gametes at isogamy are characterized by the same DNA-wave, momentum (P) and haploid number of chromosomes, but differed by a DNA-particle and spin direction. In the total pool of gametes the number of male and female gametes with spin-up and spin-down is equal. During fertilization two combinations of zygotes ♂↑+♀↓ and ♂↓+♀↑ are created.

We have revealed 5 different stages in the development of a sexual system. Naturally, the 1st step is connected with the origination of a meiosis mechanism, which has divided all sexual cells in depend of spin direction on the male and female gamete in the relation of 1:1.

The 2nd step in the evolution of the sexual process is a differentiation of gametes by the morphology and motility, and the transition of reproductive system from the isogamy to heterogamy and then from the heterogamy to oogamy. Most likely, it is caused by the fact that male and female gametes differ insignificantly by their energetics. During evolution the growth of a biological energy of a species increased the splitting of sexual cells by the level of magnetic energy. Therefore one kind of gametes in the size becomes more, and another kind less. By the conservation law of the total momentum all sexual cells should be divided into two types, because the speed increase and mass decrease of the one type of gametes should be compensated for by the loss of mobility and the rise of mass in the other type of gametes. The main evolutionary achievement at this stage consisted in the fact, that the spin direction becomes connected with a specific type of gametes - a large number of mobile and small spermatozoids (or sperms) and only one (or small number) of immovable and larger sized ovicell (or eggs).

The 3rd evolutionary step of sexual development is connected with a creation of two types of generative organs, which specialize in the formation of sexual gametes of only one specific kind. The nature of such evolutionary novelty is apparently connected with the next phenomenon. At a specific stage of differentiation the many-celled organisms create one more energy level from which there can be a transition of vegetative cells to the reproductive development. Therefore a large number of mobile and small spermatozoids or sperms begin to develop in specialized male generative organs (antheridium, different types of sporangium) and on the male generative line, which improves constantly. Accordingly, the ovicells or eggs are formed on the female generative line in another type of generative organs (oogonium, archgonium).

The male and female lines have a different stage number and structure of their matrix. The initiation of the development of the reproductive structures by a female generative line begins from a higher energy level in comparison with the male generative line.

The appearance of the cell system of Stern-Gerlach in the reproductive cycle of higher order plants is the 4th step of the sexual system development. Supposing, it were true that the given evolutionary step is connected with the creation of an activating (female) meiosis. The activating meiosis represents a combination of the two types of cell divisions – meiosis and activating mitosis. It enters into the composition of the reproductive cycle of all species of higher order plants and animals and is characterized by an identical scheme of cell divisions, spatial position, number and destiny of generative cells (Figure 34).

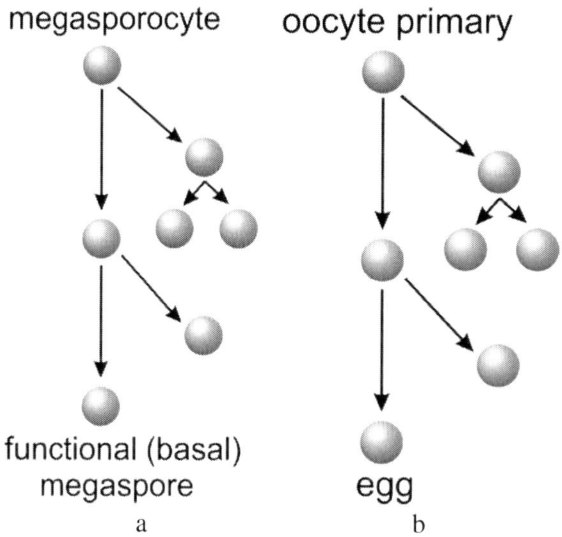

Figure 34. The scheme of divisions during activating (female) meiosis in plants (a) and animals (b).

As known, the spatial arrangement of cells in plants is more fixed concerning gravity. This gate allows it to track the changes of a spatial arrangement of generative cells during the course of two meiotic divisions. So in gymnosperm and angiosperm the activating meiosis proceeds in the center of the ovicell. During meiosis the maternal cell of the megaspore undergoes two divisions, the longitudinal axis of which coincides with a direction of gravity, and a page of megaspores is formed. Only one megaspore (functional, chalazal) is a viable gamete. It gives rise to a female gametophyte, and three other female gametes degenerate. The distinctive feature of this type of meiosis in plants is that an arrangement of megaspores in tetrads has a linear or T-shaped form.

So, we will sum up. On one hand, it has been experimentally established, that the definition of gamete sex is connected somehow with meiosis (male and female). On the other hand, in animals and plants the scheme of cell divisions during both types of meiosis completely coincides in plants and animals, up to the number of divisions of cells of the basal pole (Figure 34). Therefore for this analysis we have decided to involve one more cell division, which precedes a meiotic division of a megasporocyte. As a result, the hypothesis about the quantum-mechanical nature of sex distinctions has unexpectedly received this supplementary confirmation. We will conclude on this question with a few more details.

To start let us examine the spatial position of the generative cells during the three successive divisions - the last mitosis and two meiotic divisions on the example of the monoecious bisexual species of gymnosperms and angiosperms. In *Beta vulgaris* L. according to the cytoembryological data (Zhuzhzhalova et al., 2007) it is a generative cell and the division of a spindle oriented in a space, as shown in figure 35. Therefore a movement of the genetic material within a given part of the reproductive cycle can be represented as the movement of micro-objects in a physical system - the three "improved" apparatus of Stern-Gerlach with beams of particles with a spin of 1/2. The regularity of the behavior of particles with spin of 1/2, due to the rules of the unification of turns, has been transferred on the basis of the analyzed part of a reproductive cycle. A movement of genetic material can be represented as the transformation of micro-objects in physical system, which consists of the three "improved" apparatus of Stern-Gerlach. Each unit of this system represents a separate cell division towards the creation of a genetic program and a realization of the sexual differentiation as regards to the functional megaspores. The element S in this system connects with the formation of a maternal cell megaspore from the archesporial cell during mitotic division. The element T is the reducing division of meiosis, during which the megasporocyte formed from the dyad of the megaspores. The element U this is the second division of meiosis, which ends with the formation of a haploid chalazal (functional) megaspore. During meiosis a cell of megasporocyte undergoes two axial transverse divisions, which besides coincides with the direction of gravity. As a consequence, the systems U and T with respect to each other have the same spacial arrangement. But both of them are rotated relative to the system S on 90^0. Let us look what will occur if apparatus T and U turn together on the 90^0, when we deal with the fermions. As has been shown by the theoretical calculations and experimental data (Feynman et al., 1967), only half of the particles of one kind pass through the system.

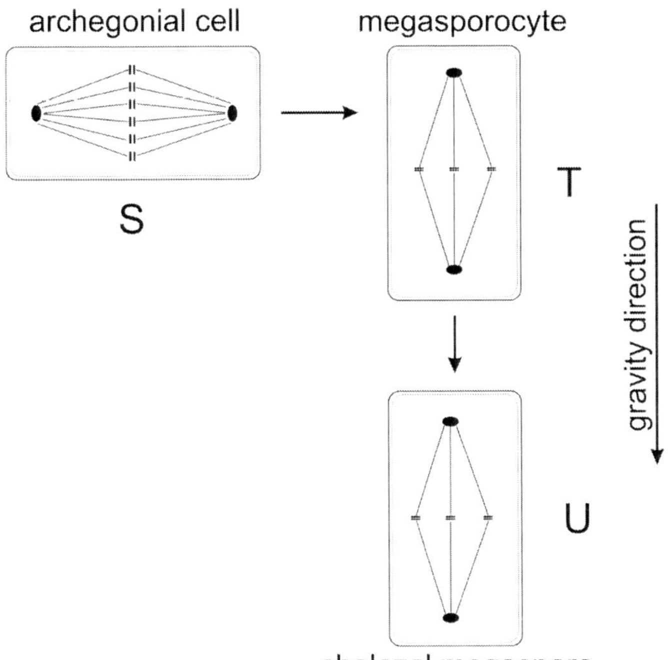

Figure 35. The scheme cell divisions of the monoecious bisexual species of plants and animals resemble the apparatus of Stern-Gerlach with beams of particles with spin 1/2.

Thus, proceeding from a spatial arrangement of metaphase spindles during three cell divisions and rules of unification of the turns, we come to the conclusion, that on the basis of one mitotic and two meiotic divisions (or apparatus of Stern-Gerlach), the definition of gametes by sex occurs by means of a non-uniform magnetic field, created by centrioles, and the filtration of set of the chromosomes (or the particles with spin of ½) in a coordinate system with a turn of 90^0 takes place. In the end of meiosis a functional megaspore is formed, which can have a haploid set of polarized chromosomes with only one direction of spin – one ↑, or one ↓. (The details of the process of the sex definition of a female gamete in depend of of the spin direction of a male gamete are described in chapter 7). The other kind of polarized chromosomes hits the nucleus of the last nonfunctional megaspore, which will soon die. Accordingly, the polarized functional megaspore gives rise to the ovicell, which has the identical direction of spin with the functional megaspore. During fertilization the ovicells with a spin ↑ will attract the sperms with a spin ↓ to itself, or the ovicells with a spin ↓, on the contrary, and the sperms with a spin ↑. Hence it follows that all species of monoecious bisexual organisms are formed with the hermaphroditic zygotes of only one type – "♂↑ + ♀↓" or "♂↓ + ♀↑".

The 5th stage in the development of the sexual system is the appearance of the phenomenon of sexual dimorphism. It should be noted, that in higher order plants the previous (bisexual) type of a sexual system is more widespread. Sexual dimorphism was found only in dioecious plants (for example, in hemp). In animals, on the contrary, it has arisen at an earlier stage of evolution and is connected with the development of sexual signs in male and female individuals. This improvement of the sexual system is the most profound. It had affected a genome of species, and, accordingly, morphology and a functioning of a vegetative and generative sphere of plants and animals at all hierarchical levels of their organization.

Now we will try to understand the nature of sexual dimorphism with an example of the reproductive cycle of the vertebrates. For this purpose we look again to the cell apparatus of Stern-Gerlah and the cytoembryological data, characterizing the spatial arrangement of generative cells during a given period of the reproductive cycle.

Oocytes originate from the cells of the germinal epithelium. We can assume, the epithelial cells undergo the activating mitosis and two energetically unequal cells are formed. The cell of the apical pole, which receives the energy (because the DNA-wave has passed to higher energy levels), later on become a central cell, and then a primary oocyte. The cell of the basal pole, which lost energy, according to the law of energy conservation continues to go down to the zero level. In this way it divides and forms a specific number of trophic cells. These cells are attracted and surround the "energetically strong" central cell, and provide it the nutrition it requires. The activating mitosis is followed by the two divisions of activating meiosis, which can proceed over many years. The meiosis leads to the formation of an egg and three polar bodies, which perish subsequently. To proceed from the particularity of the activating divisions and destiny of the generative cells, then the longitudinal axis of divisions during mitotic and two meiotic divisions should be in the same direction and coincide with the direction of gravity.

Thus, the difference between the cell apparatus of Stern-Gerlah in the monoecious and dioecious is due to the fact, that during mitosis a plane of the cell division of dioecious organisms unfolds on 90^0, after which the orientation of its spindle begins to coincide with the direction of gravity. Therefore during oogenesis it is possible for the cell divisions to present themselves in the form of three elements (Stern-Gerlah apparatus), arranged one after another along a direct line (Figure 36). The physics of the process of the filtration of gametes in this case has changed. In such a spatial arrangement of cells through this system there will pass all gametes – with spin-up and with spin-down. Accordingly, during meiosis 50% of the eggs will have a spin-up, and 50 % of the eggs will have a spin-down and under the probability theory should form within one species and within the same female individual. During fertilization 50 % of the zygotes have the combination of gametes - «♂↑ + ♀↓» and in 50 % zygotes it will be - «♂↓ + ♀↑». Note that it is the two different quantum states of the biological system. As a consequence, the origination of the dioecious individuals of one species has become inevitable.

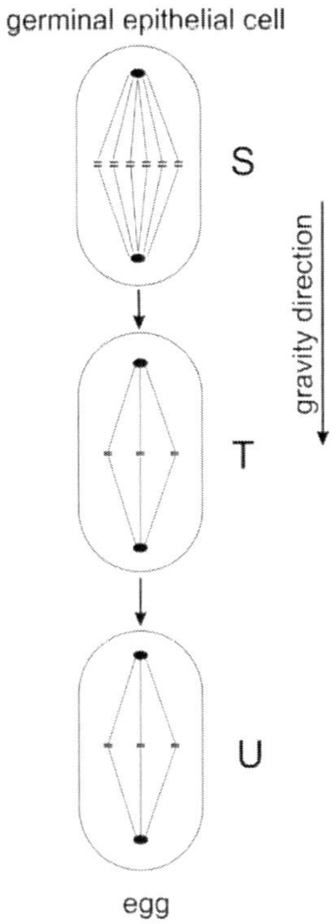

Figure 36. The spatial arrangement of generative cells during oogenesis in dioecious plants.

The last and 6th evolutionary step in the development of the sexual system concerns only the dioecious species of animals and plants. It has led to the occurrence of a distinction in the karyotype of male and female gametes and zygotes. In a set of chromosomes appears one (in this case the number of chromosomes will be odd) or two different sexual chromosomes, which are linked with a specific type of meiosis – usual, male or activating, female ones. The sexual chromosomes are distinguished by size, inner structure and function. The sex of individuals with two identical sexual chromosomes (XX), called the homogametic. The individuals with heterogametic sex have two differing chromosomes (XY), or only one sexual chromosome (X0). In most cases, the male sex is heterogametic and the female sex is homogametic (mammals, most species of insects, etc.). Now we will discuss the four types of chromosomal determinations

Drosophyla-type: females are homogametic, males are heterogametic. The mechanism of chromosomal definition of sex for the given type of sexual dimorphism is connected with a male meiosis only. In animals and plants may be two variations of zygotes of this type.

a) *Lygaeus-type.* females XX; males XY.

During meiosis the four combinations of male gametes (♂) should be, if we will proceed from the quantum-mechanical nature of sexual differences:

1) spin ↑X; 2) spin ↓X; 3) spin ↑Y and 4) spin ↓Y.

The activating meiosis produces female gametes (♀) of two types:

1) spin ↑X; and 2) spin ↓X.

Proceeding from the Pauli exclusionary principle, during fertilization two types of combinations of zygotes can appear:

1) 1♂+ 2♀; and 2♂ + I♀. Both these combinations of zygote genomes can be written, as: spin ↑↓ XX. During ontogenesis this type of zygote gives rise to the females.

2) 3♂ + 2♀; and 4♂ + 1♀, where the formula of the genome is: spin ↑↓ XY. These zygotes form the males.

b) *Protenor-type*: females are homogametic, males are heterogametic type. The females have a pair of identical sexual chromosomes (XX), and the males have a sole sexual chromosome (X0).In the Protenor-type the meiosis gives the four combinations of male gametes:

1) spin ↑X, 2) spin ↓X, 3) spin ↑0 and 4) spin ↓0.

Activating meiosis forms the two combinations of female gametes:

1) spin ↑X, and 2) spin ↓X.

During fertilization only two variations of the zygote genome by the Pauli exclusionary principle may be as well:

spin↑↓ XX - females and spin↑↓ X0 - males.

2) *Abraxas-type*: females are heterogametic types; males are homogametic types (birds, some reptiles, fish and plants). The mechanism of the chromosomal determination of sex for this type of sexual dimorphism operates during activating meiosis only. During the period of reproductive development the female individuals produce four types of ♀ gametes:

1) spin ↑Z; 2) spin ↓Z; 3) spin ↑W and 4) spin ↓W.

During another type of meiosis form the two types of gametes:

1) spin ↑Z; and 2) spin ↓Z.

The principle of formation of the genome proceeds by analogy with the *Drosophyla*-type and arise the two types of zygote combinations:

a) spin↑↓ZW – females; and spin↑↓ZZ – males;

b) spin↑↓ZO – females; and spin↑↓ZZ– males.

The appearance of sex chromosomes has changed the genome of the male and female zygotes and the male and female organisms within the same species. The differences have touched both the DNA-particle and the DNA-wave. This has provided the corpuscular-wave basis for the development of sexual and sex-linked signs separately in the male organisms and in the female organisms.

The process of the evolution of the sexual system of animals and plants continues.

3. Optical Isomery of Biological Molecules or One More Attempt to Understand the Nature of the Chirality of Life

In this chapter we want to discuss the physical model of yet one more biological phenomena. As known, the homochirality of organic compounds is a fundamental property of living matter. The fact, that in living matter there are only "left proteins" and only "right sugar" was found by Louis Pasteur in the middle of 19th century during the study of organic compounds by using polarized light (Gutina, 1990). After his work there was an obvious relationship between the optical activity of biological molecules (and the phenomenon of chirality) and a life. An optical activity was found in the carbon compounds of the rocks of the Precambrian period, which is direct evidence of a life origin. This phenomenon is inherent in all vital molecules: proteins, amino acids, nucleic acids, and carbohydrates. In cells all functioning biopolymers should be a homochiral structure, because the amino acids with the wrong chirality are capable of destabilize the helix in the proteins. Scientists are convinced that if DNA included the one monomer of the other chirality, then a macromolecule would be unable to form a polypeptide chain of the necessary length.

V.I. Vernadsky (1940) has considerably expanded the importance of chirality, emphasizing the significant role of the physical factors in the origination of this phenomenon. He wrote: "Characteristically, the compounds connected with life, prevail exclusively in one antipode"; "... the thermodynamic field of a living organism possesses a pronounced dissymmetry. We know of nothing similar among the other natural objects on the Earth". According to V.I. Vernadsky the structure of space, which is formed by living organisms, is caused by the construction of an optical isomery – a right or a left. This separates one from fundamental laws of physics, namely, Pierre Curie's principle about the initial symmetry of matter at all levels of its organization. And what is most important, is the dissymmetric developing on the Earth of "the living substance" more and more leads out from the balance of "inert matter" as between them there is a constant exchange and rotation of chemical elements.

The question of how and why the optical isomery of biological molecules is supported, if under the second law of thermodynamics it should be the racemic mixture, is the subject of constant discussion among scientists (Vernadsky, 1940; Volkenshtein, 1983; Calvin, 1971; Rutten, 1971; Bada, 1995; et. al.). Despite the fact that a number of concepts and hypotheses have been developed, and for the experimental determination of the problem not one attempt has been undertaken, however the phenomenon of homochirality so far remains a mystery of life. We want to offer a hypothesis of the origination of optical isomery in which the plasmatic membrane is represented as the special physical device – "a cellular polaroid". The plasmatic membrane is located on the path of solar (i.e., natural, non-polarized) light. In all types of cells (microorganisms, plants and animals) it has the identical ultrastructure and a typical polaroid by means of the internal organizational prompts. Below we will conclude on this question with more details.

Where a given definition of a cell, together with DNA, proteins, cytoplasm, nucleus, etc. always make mention of a cell membrane (plasmalemma, plasmatic membrane) its origination, as the chirality, goes back to the sources of life on Earth. Indeed, a cell without the external border, which ensures its integrity and independence, would not exist. The internal content of a cell differs from the external environment by the chemical compounds, energetics and osmotic pressure mainly due to the presence of a plasmatic membrane. It is characterized by the selective permeability, by means of which creates orderliness of the motion of substances into (or from) a cell. Here occurs a selection and reception of the primary information arriving from an environment, or from another cells.

The plasmatic membrane by the data of electron microscopy and roentgen diffraction represents a bimolecular lipid layer (Figure 37). The chief molecular components are phospholipids and associated with them proteins. Phospholipids - are non-spherical molecules with a length many times the width. They have an asymmetric structure: the polar heads of the molecules are hydrophilic, and the tails are neutral and hydrophobic. In a plasmatic membrane the polar ends of the molecules are turned outside, and the non-polar ends inside, forming the inner and outer hydrophilic surfaces. With the hydrophilic zone of lipids are connected different membrane proteins, which give it an electronic density. Lipids are in constant motion, which does not break the orderliness of each layer and testifies to the dynamic and liquid nature of membranes (Ivanov and Berestovsky, 1982).

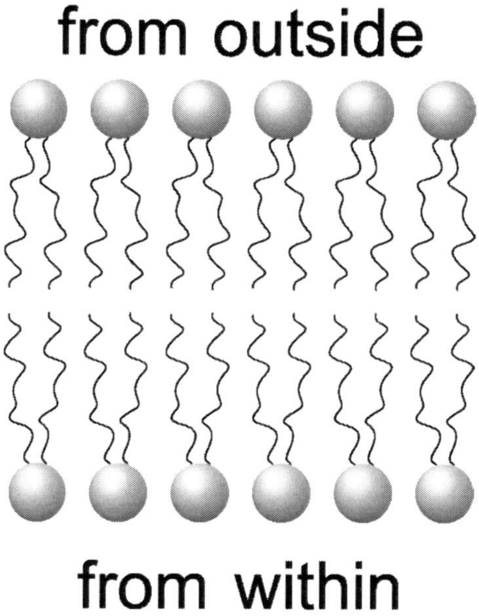

Figure 37. The scheme of an arrangement of phospholipid molecules into bilayer, which operates as the "cellular polaroid."

Another, equally important factor, on which life on the Earth depends, is the sunlight. As known, natural sunlight is un-polarized light. I.e., the direction of oscillations of the light vector is varied randomly and simultaneously in all directions, perpendicular to a ray. Therefore the question arises, whether the non-polarized sunlight can become the polarized light during its penetration into the cell, and what structure can cause a stereological homogeneity of biological molecules.

For this purpose we should consider, what occurs from a position of physics with the polarized sunlight on a boundary of two systems, when it illuminates a cell. From the total light stream we will choose one ray, which gets into a cell by the shortest optical path. The ray of light, before it can get into a cell, has to overcome the plasmatic membrane, located on its way into all types of cells. To be more exact, proceeding from the internal structure of a plasmatic membrane, then the ray in the beginning passes through the first lipid layer, which is situated on the outside of the cell, and then through the second lipid layer, which is adjoining to a cytoplasm. We assume that the natural light during passage through these two layers undergoes the following changes under the laws of geometrical optics.

As was already stated, every lipid layer of a plasmatic membrane consists of very long asymmetric molecules of phospholipids, which stretched out and positioned parallel each other. Hence it follows to the conclusion, that the external lipid layer of a plasmatic membrane represents a thin (molecular) polarizing film or a typical optical linear polarizer. The polarizer is a substance or a device, which picks out from all directions of the oscillations of natural light only one with a strictly specific direction of electric vector. Its basic property: the light, polarized along the long axis of a polarizer, freely passes through it, and absorbs the light polarized perpendicular to its axis. If un-polarized light were to pass through the polarizer, then on its exit we will receive linear-polarized light (Figure 38).

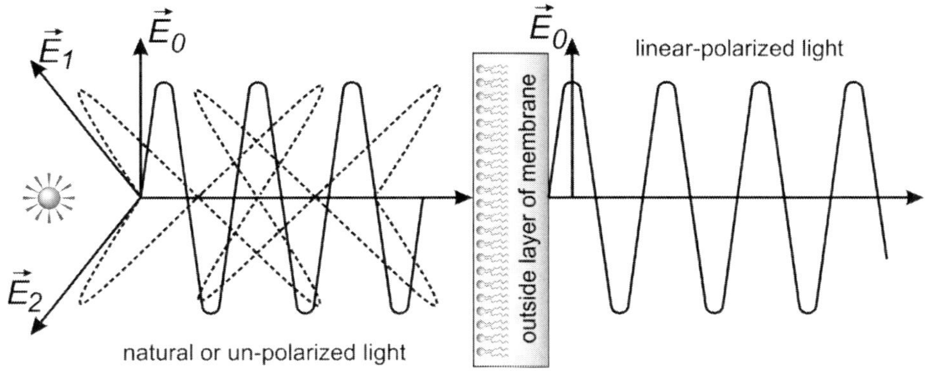

Figure 38 The principal of receipt of the linear-polarized light from a natural un-polarized sunlight during it passage through first lipid layer of plasmatic membrane.

Thus, when the sunlight passes through the external lipid layer of a plasmatic membrane (or through a molecular polarizing film) the ray comes out of it by a linearly polarized light. After that the linearly polarized light enters into the second lipid layer. Because the molecules phospholipids do not possess mirror symmetry, during the passage of polarized light through them under the law of physics, will cause one more polarizing effect, which is known as the phenomenon of optical activity.

As known, the phospholipids are asymmetric molecules. Each of them from the side of the tail looks like a corkscrew. Therefore the lipid bilayer is composed of many molecules in the form of a spiral which are located parallel to each other. When the linearly polarized light gets on a spiral molecule, then the electric field inside it begins to induce a motion of charges upwards and downwards by a spiral that conducts to the resulting field with the appearance of displacement. The direction of the plane of polarization of linearly polarized light is rotated around the axis of the ray, giving the effect of optical activity.

From the above follows an important conclusion. The reason for the chirality of biological systems can be a plasmatic membrane. Its physical and chemical structure represents the two molecular polarizing films (or two polaroids), consisting from a long asymmetric identically oriented molecule of phospholipids. The first lipid monolayer of the plasmatic membrane transforms the un-polarized ray of light to a plane-polarized ray, and then the plane-polarized ray during its passage through the second monolayer is rotated around its own axis and the polarized ray on a circle of light enters into a cell. It is considered, that an optical active substance can be created only by another optical active substance (Rutten, 1971). Therefore a circularly polarized light, as the optical active substance, can cause the optical activity of biogenic compounds.

According to our notions, the cell without a biological harmonious oscillator cannot exist. In prokaryotic cells the oscillations of a DNA-wave (or a RNA-wave) are limited outside of a plasmatic membrane. Its lipid bilayer, as the two molecular polarizing films, is capable of converting a natural sunlight into the circularly polarized light.

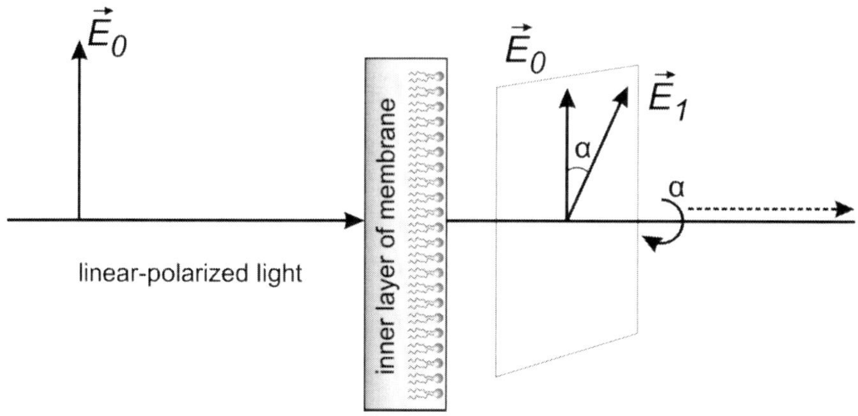

Figure 39. The transformation of linearly-polarized light into circularly polarized light.

It is experimentally established, that a circularly polarized light is absorbed differently by the left and right stereoisomers and causes in them the effect of destruction by light. Therefore we can assume, that at the beginning of evolution the circularly polarized light, getting on the racemic mixture of dissymmetrical molecules, gave only one of them an advantage, creating in the cells metabolic conditions on the basis of one kind of the stereo-isomers of enzymes, proteins, carbohydrates, and other biomolecules. A unidirectional selection led to the survival of organic molecules of one sort and an origination of optical isomery, which during evolution was firmly fixed in living matter. As a consequence, all biomolecules in a cell are found not as a mixture of stereo-isomers, but only in one of its forms - L or D (the majority of amino acids is in the L-form, almost all sugar in the D-form, etc.).

Conclusion

In the given chapter we have tried to show the leading role of physics laws on examples of different biological processes. The quantum-mechanical nature of a cell opens a huge sphere of activity for many investigations. However a DNA-wave is the energy, a very great energy. The following chapter is devoted to the questions, of how nature "reasonably" uses the energy of a DNA-wave and during sexual reproduction is protected from its negative action.

References

Alberts, B., Johnson, A., Lewis, J., Raff, M., Roberts, K. and Walter, P. (2002). *Molecular biology of the cell*, 4th edn. Chaffey, N (Ed.), Garland Science, New York.

Bada, J.L. (1995). Origins of homochirality.*Nature*, 374 (6523), 594-595.

Belousov, L.V. (1980). *Introduction to general embryology*. Moscow, Moscow State University.

Belousov, L.V. (2005). *Bases of general embryology*. Moscow, Moscow State University.

Bogdanov, Yu. F. (1991). The homologous conjugation of chromosomes. *Improvements of recent genetics*. pp. 61-134. Moscow, Nauka.

Calvin, M. (1971). *Chemical evolution*. M: Mir.

Chentsov, Yu.S. (2004). *Introduction to cell biology*. Akademkniga, Moscow.

Cytology and genetic of meiosis. (1975). Eds. V.V. Hvostova and Yu.F. Bogdanov, Moscow, Nauka.

Darlington, C.D. (1965). *Cytology*. J. and A. Churchill LDT, London.

Gutina, V.N. (1990). L. Pasteur and V.I. Vernadsky about a life origin. *Questions of history of the natural sciences and technique*, 2, 36-44.

Embryology of Angiosperms. (1984). Ed. B.V. Johri. Springer Verlag. Berlin, Heidelborg, New York, Tokyo.

Feynman, R.P., Leighton, R.B. and Sands, M. (1963). *The Feynman Lectures of Physics*. vol. 1-3. Wesley, A. (Ed.), Readigg, Massachusetts, Palo, Alto, London.

Gilbert, S.F. (2010). *Developmental biology*. 9th ed., Sinauer Associates, Inc., Sunderland, Massachusetts.

Ivanov, V. G. and Berestovsky, T. N. (1982). Lipid bilayer of biological membranes. M: Nauka.

John, B. (1990). *Meiosis*. Developmental and Cell Biology, Ser. 22, Melburn, Cambridge: Cambridge Univ. Press.

Komarnitskii, N.A., Kudryashov, L.V. and Uranjv, L.V. (1975). *Systematic of plants*. Moscow, Nauka.

Kozubov, G.M., Trenin, V.V., Tihova, M.A. and Kondratieva, V.P.(1982). *Reproductive structures of gymnospermae (comparative description)*. Leningrad, Nauka.

Life of plant. (1978). V. 1-6. I.V. Grushevitskii and S.G. Zilin eds. Moscow, Prosveshenie.

Petrov, D.F. (1973). *The cytological basis of heredity*. Moscow, Kolos.

Poddubnaya-Arnoldi, V.A. (1976). *Cytoembryology of angiosperms plant. Principles and perspectives*. Moscow, Nauka.

Reyvn, P., Ewert, R. and Aykhorn, C. (1990). *Modern botany*. V. 1, Moscow, Mir.

Rutten, M.G. (1971). *The origin of life by natural causes*. Elsevier Publ. Comp. Amsterdam, London, New York.

Singh, H. (1978). Embryology of Gymnosperms. *Encyclopedia of plant anatomy X (2)*. Berlin, Stuttgart, Gebruder Borntraeger.

Swanson, C.P. and Webster, P.L. (1977). *The cell*, 4th edn. Prentice-Hall Inc., Englewood, Cliffs, Jersey.

Sybenga, J. (1975). *Meiotic configurations*. Springer-Verlag, Berlin, Heidelberg, New York.

Tokin, B.P. (1987). *General embryology*. 4th ed., Moscow, Visshaya shkola.

Vasiliev, A.E, Voronin, N.S. and Yelenevskii, A.G., etc. (1988). *Botany.Morphology and anatomy of plants*. 2nd ed. Moscow, Prosvechenie.

Vernadsky, V. I. (1940). Studying of phenomena of a life and the new physics. *Biogeochemical essays*. M: AN USSR. P. 175-197.

Volkenshtein, M.V. (1983). *Theoretical biophysics*. Nauka, Moscow.

Zhuzhzhalova T.P., Znamensky V.V., Podvigina O. A, Jarmoljuk G. I. (2007). *Reproductive biology of sugar beet*. Voronezh: OOO «Sotrudnechestvo».

Chapter VI

Energy Protection System of Plants: Mechanisms, Ways and Evolution

Abstract

The wave nature of generative cells makes essential alterations in the organization of the sexual process of plants and animals. The special system of energy protection for the regulation of interactions between somatic and generative cells and tissues is created. In the result of deciphering of the reproductive cycle was established, that all stages are divided on three categories dependent on functional predestination: 1) the stages, which participate in an "bioassemblage" of corpuscular (hereditary) program and 2) the energy program of the development of the genotype of a future plant; 3) the stages by which the formation of new reproduction initiates, as well as 4) the stages, which limit the growth of generative cells on a sporophyte by a fixed volume (Kuznetsova, 2002). In this chapter is examined the last type of stages or the problem of the energy protection of a sporophyte. There are the processes, which take place in a contact zone between somatic and generative cells and tissues (for example, the female gametophyte and nucellus), or between the two generative cells with different quantities of biological energy (embryo and endosperm) during their simultaneous development on the body of a maternal plant.

The quantity of biological energy in somatic and generative cells constantly varies. It is the peculiarity of the work of a biological harmonic oscillator as a physical system. Nevertheless these cell systems depend on the energy state of their genome so it is possible conditionally to divide the two categories of cells into – "strong" and "weak". This has allowed us to analyze the processes of incompatibility between the cells which differ by a DNA-particle and/or by a DNA-wave. All the ways which during evolution were created for the protection of sporophyte from the influence of the high energy of generative cells, we will describe as the system of "energy protection". Revealed the two mechanisms (a removal of reproductive structures from a maternal plant and a transition of generative cells in a resting state) and the four ways (haplont way, external fertilization, bimorph way and double protection) of energy protection. The change of the character of relations between vegetative and generative cells during the process of sexual reproduction; the peculiarities of the formation of separating structures in a contact zone and their significance; the role of the genetic system of incompatibility in reproductive processes; the pathways of the evolutionary development of an energy protection system in plants are discussed.

Vihrov in 1858 had defined the essence of cell theory as "*Omnis cellula a cellula*" (Huger, 1959; Zavarzin et al., 1992; Chentsov, 2004). Then the cytologists in the 20th century have come to the conclusion, that the daughter cells during mitosis receive a full set of chromosomes and the number of them does not vary during an ontogeny. According to modern cell theory the genetic program of an individual is formed on the basis of a two-chained DNA molecule (i.e., on the basis of the chemical formula of a DNA-particle) and connected with the specific sequences of nucleotides. A nucleus of differentiated cells has the same genetic program, as that of the nucleus of a zygote. The given conclusion has been confirmed by numerous experiences of the transplantation of nuclei (Brachet, 1960; Tokin, 1987; Belousov, 2005; Gilbert, 2010).

Another situation also takes place in embryology (Korotkova and Tokin. 1977; Tokin, 1987; Belousov, 2005; Gilbert, 2010; etc.). Till now the experimental data cannot be explained within the modern knowledge of embryology and developmental biology. Beginning from Aristotle the questions of reconstruction and loss of totipotency, the orderliness and disorder of cell systems, etc. which underlie in the basis of all researches and theories, is devoted to the developmental biology of animals and plants (preformism, epigenesis, continuity of germinal plasm, embryonic fields, physiological gradients, etc.). Many theoretical statements and hypotheses are disputable and serve as a subject of constant discussion. The absence of settled concepts, possibly, is caused by the nature of embryology as the object of research is more closely connected with the manifestation of wave properties of living matter, than with the corpuscular properties.

Accidents in nature do not exist and everything is developing on the basis of regularities and laws. Berg (1977) wrote: "A life, we will tell, - is like a clock with an unusually prolonged, maybe, eternal winding mechanism: being once wound at the daybreak of life history, this clock continues to keep a store of energy, transferring it from generation to generation". The biological harmonious oscillator or DNA-wave appears as such "a clock" in our corpuscular-wave model. From the moment of the origination "at the daybreak of life history" of a harmonic oscillator in the ancient cells and up to present day in all types of cells of different biosystems are not damping oscillations – the auto-oscillations are excited and are sustained from generation to generation and from species to species. Accordingly, unicellular and multicellular organisms have a hereditary (on the basis of a DNA-particle) and energy (on the basis of a DNA-wave) programs of development, which independently from each other are laid out and are realized during the cell cycle and during the period of sexual reproduction and ontogenesis. During evolution both these programs constantly improve at different levels of the organization of the species - from molecular to specific. The progress of the speciation process on a DNA-wave base depends mainly on the degree of development of a generative sphere of species and the energy capacity of the harmonic oscillator of the last generative cell (zygote, sporophyte initial). It is very large amount of energy. That is why the evolutionary development of living matter should be going in two directions.

1) An increase of capacity of a biological harmonious oscillator and energy potential of a species, which are functioning at the expense of the directional change of a DNA-particle (increase of chromosome numbers) and a DNA-wave (transition of DNA-wave to higher energy levels). The given questions in more details will be discussed in the following chapters.

2) A perfection of the system of energy protection, which regulate the relations between the generative and vegetative sphere of a plant. In this chapter the next questions will be considered: the correlation between the wave properties of generative cells and the organization of the reproductive process; the reasons for the periodic change of means of energy protection; and the role of a given system for the evolution of plants.

Energetically Strong Cells

If we are to proceed from the corpuscular-wave nature of DNA, then the transition of the DNA-wave to the higher energy levels presents the direction for the rejuvenation of a genome in one or a few generative cells. Living matter has mastered it during evolution on the basis of the wave nature of the DNA molecule. It is a genetically determined many-stage process, providing a preservation of wave information of the species during alternation of sexual generations. The wave development of generative cells is carried out on a matrix of the energy line of the reproductive cycle, the structure of which is characterized by a high species-specificity.

On the maternal plant the specific physical conditions (intracellular pressure, polarity of generative cells, etc.) should be secured during creation of the energy program in a future plant. In eukaryotic organisms for this purpose the mechanism of energy transfer from the one chromosome to another during special type of mitosis (activating mitosis) is used. The generative cell, which has received energy during activating mitosis, becomes an energetically strong cell. An energetically strong cell acquires the properties, which distinguish it, first, from its predecessors of generative cells and, second, from the neighboring cells and tissues. In this case potencies in the developmental and metabolic activity are increased. The ultrastructure of the cytoplasm as a result of rejuvenation is changed also. All the other cells (generative cells which lost energy during activating mitosis, and neighboring vegetative cells), which we will name the energetically weak cells, at this time continue to divide and aspire to the equilibrium (or to the energy of zero oscillations) according to the second law of thermodynamics. However now, a gradient appears between the condition of a DNA-wave in the energetically strong cells and energetically weak ones. It leads to violations and death of energetically weak cells in the zone of their contact. It is necessary to notice, that the division of cells on the energetically strong and weak cells is conditional. For example, the cells of the female gametophyte are the energetically strong ones in regard to a nucellus tissue and at the same time are the energetically weak ones in regard to an embryo. Therefore it is necessary to take into consideration the vector direction of the energy in a zone of contact between the energetically strong and weak cells.

As is known, the processes of seed formation in higher order plants proceeds on the body of a sporophyte. Therefore a situation during sexual reproduction is periodically created, when the two groups of cells with a different capacity of a harmonic oscillator simultaneously co-exist on a body of a maternal plant: female gametophyte – nucellus; sporogenous cells – microsporangium wall; pollen tube - nucellus; ovicell – female gametophyte; etc. The energetically strong generative cells after origin begin to develop on a body of a maternal plant. It can occur in the following ways: an increase of cell volume (ovicell); an intensive crushing of nuclei without a cytokinesis or with the subsequent cytokinesis (embryo sac, female gametophyte, embryo in gymnosperm); a simultaneous increase of the size and the number of cells (archesporium). In the beginning the growth of non-differentiated generative tissue goes exponentially. And if there were no constraining mechanisms from the side of energetically weak cells (somatic cells maternal plant, tapetum tissue, etc.), then this non-differentiated mass of cells would continue to grow as “a cancerous tumor” up to the death of the plant.

Energetically Weak Cells

The energetically strong cells are always the generative and the strongest among them is the last generative cell of reproductive cycle (meiospore, zygote, etc.). The energetically weak cells can be the generative and vegetative ones and the weakest are all equilibrium cells (synergids, prothallial cells, etc.). After the origination of new reproductive structures, the nearest energetically weak cells react to their appearance in the identical manner. They turn to them. A metabolism, the longitudinal axis of symmetry and the planes of divisions are changed. For example, the response of a nucellus on the origination of a functional megaspore is the appearance of a nucellar tapetum (Figure 40-a); the response of a female gametophyte on the origination of an ovicell is the appearance of jacket cells of arhegonium; etc. Natural, the spatial arrangement of the cells of the nucellus and endosperm could not be altered by themselves. Consequently, some force has been applied, which has pulled them from their original places, turned and has arrange them in radial rows around a new reproductive structure. It is possible to explain such a behavior of the cells of a nucellus and endosperm in a contact zone if you take into consideration that we are dealing with the oscillatory systems.

At the transition of the DNA-wave to higher energy levels the value (intensity and extent) of the biofield is increased. The tension of the field strength is the highest level in the zone around the nucleus. In the far zone it decreases to 1/r as far as the removal of a generator. Therefore energetically weak cells, which have reached a zone of action of a more powerful harmonic oscillator, react to it and result in a motion. Besides the displacement, acceleration and speed of motion are directly proportional to the tension of a biofield.

The regimen of capture proceeding from the wave nature of the process begins to operate in a contact zone between the energetically strong and weak cells. This physical phenomenon is created by a harmonious oscillator of energetically strong reproductive structure in regards to the specific number of the harmonic oscillators of energetically weak cells. Their DNA-wave under its influence are passed from an auto-oscillation to a beat, and then from a beat to a frequency of oscillation of the energetically strong structure. Such cells "forget" their own program of development and are under the influence of another, stronger program. This has changed their destiny. The increase of the oscillation frequency of a DNA-wave induces the essential shift of a metabolism of cells and an acceleration of synthetic processes contrary to its own program. Therefore in a zone of contact the energetically weak cells are transformed into the cells of a secretory type. The secretory cells are characterized by a specific ultrastructure and high synthetic activity. This type of cell is typical for a tapetum of microsporangium and nucellus, jacket cells, etc. (Kapil and Tiwari, 1978; Konar and Moitra, 1980; Willemse and Boer-Dejev, 1981; Kozubov et al., 1982; Trenin, 1986, 1988;

Kuznetsova, 1990; Pacini, 1997; Chehregani, 2008; etc.). They consist of the darkly stained cells with a large number of organelles. The endoplasmic reticulum, mitochondrion and plastids often occupy the majority of its volume (Figure 40-b,c). The new synthesized biomolecules are not used for their own metabolism of secretory cells, in spite of the fact that have a high synthetic activity. All nutrients begin to move by the vector of a stronger biofield and are used for the nutrition of quickly growing reproductive structures. The intensive operating program is harmful for the secretory cells. The processes of lysis and signs of degeneration have appeared, are accelerated and lead them to death. The cells of this layer are squeezed and intracellular contents are absorbed by the reproductive structure. Then the cells from the next layer occupy their place. In a zone of contact, they get into a region of high tension in the biofield, are transformed into the secretory cells and then perish. The process of the rotation of energetically weak cells and the growth of reproductive structures on a maternal plant, will be continued until then the energy gradient between the two cell systems will remain on the same level. From our point of view, the gradient between somatic and generative tissue is great enough and during cell divisions does not decrease, so far as the speed of radiation of a biological harmonious oscillator is identical in both cell types - energetically strong and energetically weak.

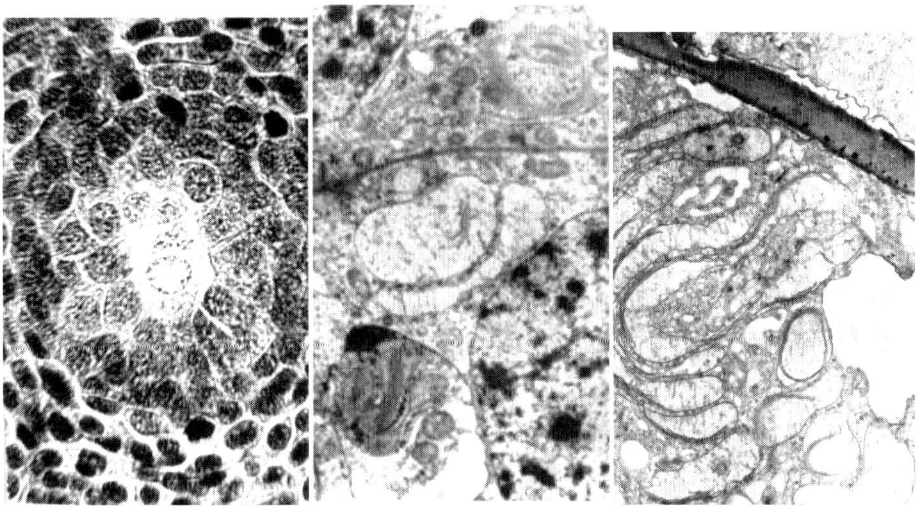

a b c

Figure 40. Nucellar tapetum during the stage of megaspore tetrad (a, courtesy of Dr. V.S. Svintsova) and ultrastructure of the jacket cells of the archegonium (b, c) in *Pinus sylvestris* L.

Mechanisms of Energy Protection

Formation of the Abscission Layer

This mechanism of energy protection acts on the basis the processes of formation of abscission layer on the boundary between energetically strong and energetically weak cell systems. The generative cells, which have lifted their own level of biological energy, are passing to the following stage of the reproductive cycle. The continuous receipt of nutrients is necessary for the further development of new generative cells and structures. The trophical function is provided by means of nearby nutritious tissues, whose cells enter into a program of high synthetic activity and a rotation in conformity with the scheme described above. However all nutrients before utilization should pass over a dividing barrier between the two cell systems. It can be a plasmatic membrane of one energetically strong cell (envelope of microsporocyte), or a double barrier: the plasmatic membranes of individual cells together with the common envelope covering all reproductive structure outside in the form of a sheath (sporoderm of female gametophyte).

As known the plasmatic membranes possess a selective permeability. Therefore one kind of molecule easily passes through this barrier, another with difficulty, and for some molecules this barrier will be an unsurmountable obstacle (Clarckson, 1974; Ling, 1984; Lehninger, 1985; Rubin, 1999). The plasmatic membrane does not limit a diffusion of small molecules: water, carbonic dioxide, oxygen, etc. A passive transport through a membrane moves spontaneously in accordance with a concentration gradient. Receipt in a cell of larger molecules, such as a sugars, amino acids, nucleosides, etc. occurs by means of a facilitated diffusion. The transfer of molecules (ions) against an electrochemical gradient is carried out by means of active transport and requires an expenditure of energy. As a consequence a non-equilibrium distribution of ions (K^+, Na^+, H^+, Ca^{++}, Cl^-, etc.) is achieved within the cells and between the cell and the environment. Undoubtedly, when a cell is alive, the forces of a biofield should take part in the processes of realization and regulation of these two types of transport. This is confirmed by the fact that after the transition of a cell from an open system to a closed system (mitosis beginning) the motion of molecules and ions that occur due to the electrochemical gradient and the concentration of ions inside and outside the cells begin to even out (Clarckson, 1974; Sorokina, 1978; Kazmin, 1984). Molecules with a large molecular weight (polysaccharides, some proteins, complicated lipids, etc.) are not capable of getting through this dividing barrier. The plasmatic membrane sorts the molecules of a chemical compound from a stream, which moves from the numerous energetically weak cells to the new energetically strong structure. As a consequence, not all molecules and ions, moving from the energetically weak cells will be used for its metabolism. Some of them are detained and settle on the outer side of the membrane (or common boundary) and progressively accumulate around it. This region is the zone, where biofield tension is rather great. Probably therefore the sedimentary identical biomolecules, densely pressed into each other by means of biofield forces, begin to polymerize. Some authors (Swanson and Webster, 1977) compare this process of polymerization with the production of plastic or synthetic fibers.

It seems obvious, that the special formation is developed around all energetically strong structures, which we call the abscission layer. A sporoderm of pollen and female gametophytes, an envelope of an ovicell, microspore, etc. all of these are the abscission layers (Figure 41). These structures are the main element in a system of energy protection and play an important role in the processes of sexual reproduction. They differ by form, thickness and consist mostly of the high-molecular polymers such as callose, sporopollenin, cellulose, etc., which according to the research have received the label of extracellular substances. For example, the seed coat, which in most cases of angiosperms consists of a firm outer layer, is a reliable protection for the embryo within the seed (Villiers, 1972; Esau, 1977). It regulates seed germination according to time and on the dependency of the weather conditions, because of the low permeability to water. The seed coat preserves the seeds not only from adverse weather conditions, but also promotes a distribution of seeds into the different habitats.

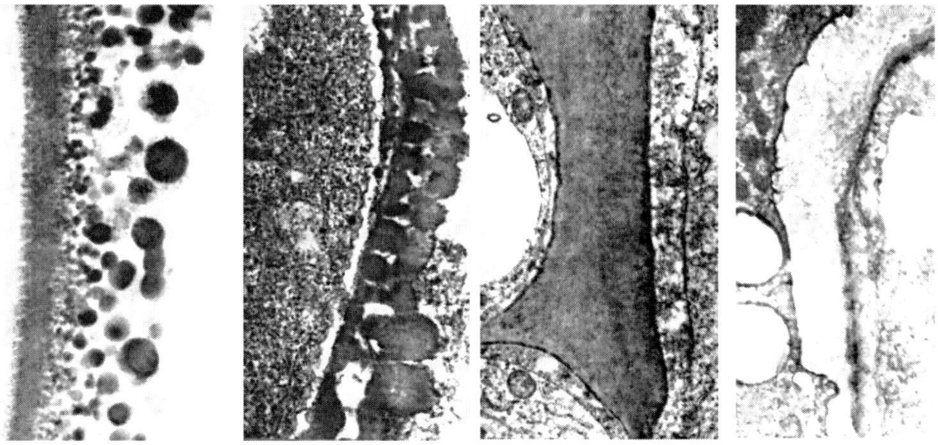

a b c d

Figure 41. Abscission layer between energy strong and energy weak cells in zone of contact: the sporoderm of female gametophyte (a) and pollen (b); envelope of ovicell (c) and pollen tube (d).

The extracellular substances are characterized by a high stability, the insolubility in water and the carrying out of a number of vital functions during sexual reproduction and dispersal of seeds (Swanson and Webster, 1977). Proteins, entering into their composition, have an influence on the surface properties of the cells. It gives them the ability of mutual "recognition". Most likely the extracellular substances have another important property - the effect of an insulator, or the ability to reduce (extinguish) the tension of an electromagnetic field, which emanates from the energetically strong cells, up to the level of the surrounding tissue. The high stability of these polymeric substances holds back the growth of reproductive structures inside the specific fixed volume. All these properties of an abscission layer provide the reliable protection both of the sporophyte and the reproductive structures.

The mechanism of the formation of an abscission layer is connected with the wave nature of the cell and is the same in all species of higher order plants. The energetically strong generative cell originates after an activating mitosis (or activating meiosis) and as rule has the usual cell envelope. The abscission layer begins to form from the moment of it appearance and proceeds during the whole period of the joint development of two cell systems with a different energy potential. A powerful biofield initiates the process of its development. This field creates a centripetal stream of nutrients through a plasmatic membrane. When the thickness of an envelope is small and is capable of exercising its influence on the growth of generative tissue is insignificant. Then during a progressive thickening it begins to hamper the growth of the generative tissue. After reaching a threshold concentration the given envelope becomes like an "impenetrable case" inside of which the generative cells are placed. The cells up to this time are in a state of rest. Every abscission layer (sporoderm of pollen grain, sporoderm of female gametophyte, callose envelopes of microspore, megaspore and ovicell) is characterized by a species-specific chemical compound, thickness and often consists of a few layers, which differs by origin.

Now we will examine the course of events during the formation of the abscission layer, but only from the position of the energetically strong cells and their derivatives. As was mentioned above, the generative cell after transition to the higher levels of energy acquires the additional energy potential, which begins to realize on a body of maternal plant. Without a durable abscission layer stopping this powerful proliferous process is impossible. The restriction stage is the period during which the abscission layer should be formed and generative cells should pass into a state of physiological rest. This process is due to several stages: 1) an exponential growth; 2) a transition from a stage of logarithmic growth to the stationary phase; 3) an internal differentiation; and 4) a physiological rest. The ability of a cell to cross into a dormant state is regarded as one of the fundamental properties of cell systems (Lajtha, 1963; Epifanova, and Terskich, 1968). The period of rest is a specific state of a cell, when growth and division are stopped, and the intensity of the metabolism sharply decreases but the viability and potentialities to further development of the generative cells is not lost. The ability of cells to survive in an adverse environment is increased in this state as well. The cells are capable out of leaving a dormant state only under specific conditions (internal and external) and again start the cell divisions.

A restriction stage of the development of the reproductive structures is the genetically determinate process. The main requirement for its realization during the reproductive process is that the stages should change each other gradually. In this case the transition will occur without an injury to the genetic material. The optimum program of the proceeding method of restriction is provided by the following factors. A constant thickening of the insulating envelope reduces the stream of nutrients into a cell. Its ability to stretch also decreases gradually. The last factor leads to an increase of intracellular pressure for the following reason.

Any cell is primordially characterized by a specific volume and species-specific intracellular pressure. Each division is completed by a cell doubling which by volume becomes two times more than the initial cell. As soon as the insulating envelope reaches a thickness, at which it is able to limit the growth of the reproductive structures, then by the gas laws this should lead to an increase of intracellular pressure. On the other side a DNA-wave radiation as free energy source is the periodic physical phenomenon, when cell divisions takes place. At the optimal regime of development of the abscission layer and restraining stage the generative cell has time for the correction of its own metabolism and development, which should correspond to the new physical conditions. So after origination the nucleus of a functional megaspore in *Pinus sylvestris* is divided actively and a female gametophyte quickly increases in volume. Then the proliferous process is stopped and the nuclear gametophyte passes to the stage of internal differentiation and gives rise to the endosperm tissue, archegonial initial, ovicell, etc. After that, an accumulation of the supply of nutrients in the endosperm cells is beginning. At the end of endospermogenesis a DNA-wave radiation stops and the cells pass into a rest state. The intracellular structure of the endosperm cells is transformed in accordance with the stages of restraint. In figure 19-a and b it is shown that their ultrastructure during a restraining period has changed. At the beginning of the cellular stage it is a strongly vacuolated cell with a thin parietal layer of cytoplasm. Then the volume of cytoplasm and number of organelles are increased and the cells are filled by lipids and aleuronic grains.

Let us analyze the nature of a restraining stage from the position of physics. As known all processes according to the second law of thermodynamics should proceed in the direction of equilibrium. The considerable increase of the energy level in the part of the generative cells leads a reproductive system out of an equilibrium state. At first this non-equilibrium system consists of one or several energetically strong cells and a large number of energetically weak cells (functional megaspore – nucellus; central cell – endosperm; tier of embryonic cells – endosperm; etc.), between which the semipermeable plasmatic membrane is located. The new reproductive structure quickly grows, and at this time the abscission layer on the boundary of the two tissues is formed. The envelope during thickening gradually reduces the intensity of the biofield around the energetically strong reproductive structures to the level of the neighboring somatic or generative tissue, stops their growth and translates the generative cells into a rest state. Therefore the restraining stage is considered as the process, which intends to restore the lost balance. As a consequence the reproductive system comes back again in a condition of the average distribution of energy to the end of the restraining stage of its development on a maternal plant. In equilibrium reproductive structures are not capable of doing work and cannot leave the rest state without assistance. All of them are doomed to death, if they remain in a former position and in a former quality. That is why non-fertilized ovicells and non-germinated seeds always perish.

The two ways that reproductive structures can leave a rest state are as follows. 1. That part of the reproductive structures (pollen, seeds) leave a maternal plant and pass to a new habitat, where a further development of them is renewed. For example, the pollen is transferred into the pollen chamber (gymnosperms) or onto the stigma (angiosperms). The new inhabitancy pushes the generative cells from a rest state and provides a growth of the pollen tube by the nutrients, which accumulate in the cells of a nucellus, stigma and style. 2. Other part of the structures (microspore, ovicell) will continue their development on a maternal plant by means of passage on to the next stage of the reproductive cycle. The exit of generative cells from a rest state can occur in these two ways: 1) the destruction from within of the old abscission layer and the formation of a stronger barrier structure (microspore); 2) the formation of a new abscission layer and the preservation of an old envelope (female gametophyte). For example, a microspore transition in a stage of microgametophytogenesis is accompanied by a destruction of the callose envelopes and a formation around the pollen grains of sporoderm, which contain a sporopollenin. This more stable abscission layer begins to form, when the tetrads of a microspore still conclude inside of a callose layer (Mepham and Lane, 1970; Ezau, 1977). Most likely, one of the reasons of callose layer destruction is radiation from a DNA-wave. Its energy initiates the mitotic division of a microspore in a microsporangium, causing a growth of intracellular pressure and a destruction of the callose envelope from within.

During sexual reproduction the genome of generative cells is formed on the basis of both a DNA-particle and a DNA-wave, because the reproductive cycle of all species of animals and plants includes such types of cell division as the meiosis and activating mitosis. The energy potential of DNA-wave changes during activating mitosis. During meiotic divisions a number of chromosomes decrease by half, and the male and female gametes receive a different DNA-particle. Its chemical formula is changed by means of the special intracellular mechanism – a crossing-over (see Chapter 3). Therefore the boundary conditions between the genomes of the two cell systems can have different combinations dependent on the stage of reproductive cycle: vegetative and generative cells, haploid and diploid cells, energetically strong and energetically weak cells, genetically homogenous and genetically heterogeneous cells. When such cells and tissues begin to interact the abscission layer between them is formed. The speed of formation, chemical compound and thickness of the envelope depend considerably on the intensity of these incompatibility reactions.

A genetic system of incompatibility demonstrates the complex biochemical and immunologic reactions between the two heterogeneous cell groups and tissues (Heslop-Harrison, 1978, 1983; Pandey, 1980; Zavada and Taylor, 1986; Dickinson, 1990; Vishnyakova, 1997). Incompatibility as a biological phenomenon regulates a close relationship of species in a system of "inbreeding – outbreeding", provides the interspecific reproductive isolation at the time of pollination and the plants survival in an adverse environment. During the course of a normal reproductive process its reactions bring the essential contribution to the processes of the formation of an abscission layer. In higher order plants in the contact zone between the two cell systems, which differ by a DNA-particle and DNA-wave, the callose substance is often revealed. It is the polysaccharide that is insoluble in water and consisting from the rest of the glucose molecules, which connect in a spiral chain by means of P-1, 3 bonds. A callose enters the composition of a microspore, megaspore, ovicell and pollen tube envelopes. The callose layer reaches a considerable thickness around the microspores and does not have plasmodesms. It testifies to the intensity of the incompatibility reactions, which take place between the haploid microspores and diploid cells of anther wall (Figure 42).

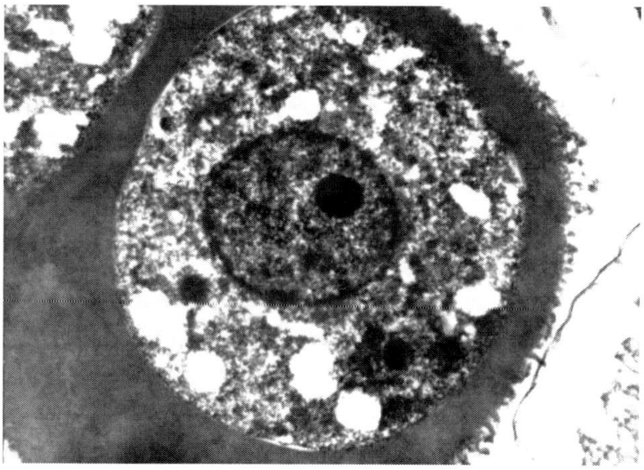

Figure 42. Callose envelope of the microspore of *Quercus robur*L.

As was mention above each stage of energy protection is a genetically determined process. Therefore all reproductive structures (microspore, pollen, ovicell, endosperm and seed) after a restrained stage of development have a species-specific morphology and size (Buchhollz, 1946; Singh, 1978). In a microsporangium the volume of sporogenous tissue and the reserve of vegetative cells, which are necessary for nourishment of the developing microspores and pollen, are carefully connected to each other and adjusted. Really, all layers of a microsporangium wall with the exception the last epidermal layer are spent on the trophic function of pollen during its formation. A destruction of the epidermal layer provides the exit for the pollen and its participation in pollination. The tissues of the nucellus and integument contain exactly the quantity of somatic cells and nutrients, which for the restraining stage of the development of a functional megaspore during megagametophytogenesis are necessary.

A relation between the volume of the endosperm and the embryo in seeds also is defined for every species. The endosperm also contains a specific quantity of cells and nutrients, which will be used during embryogenesis and seed germination. Concerning the quantity of endosperm into a seed, its volume, apparently, is connected with the value of the energy gradient between the tissues of the embryo and endosperm and, accordingly, with the length of the restraining stage during embryogenesis.

Separation of the Reproductive Cycle in a Space

This mechanism of energy protection acts on the basis of the removal of part of the generative cells from the body of a maternal plant before the process of their reproductive development will complete. A separation of a reproductive cycle on the two independent living forms (sporophyte and prothallium) becomes possible when the reproductive cycle of the species has a certain number of stages. It should be noted, that during evolution the reproductive cycle becomes longer, mainly, at the expense of the energy line. For example, the number of activation stages, on which take place a lifting of the DNA-wave to higher levels of energy, has increased from one (*Conjugatophyceae*) to six (Angiospermae). During the period of sexual reproduction the energy line operates in this manner, that a quantity of biological energy in each subsequent generative cell has a higher level than in the previous ones. Therefore the energy potential of a zygote is always above that which is in a functional megaspore. The apical point of a reproductive cycle is the last generative cell (zygote, meiospore, initial sporophyte), which possesses the maximum biological energy.

The rupture of a reproductive cycle and the removal of its energetically strong generative cell (or reproductive structures) from a body of plant is the effective mechanism of protection of maternal plant, which has received a wide circulation in vegetative and zoological organisms. In plants the given mechanism meets for the first time in the *Fucus* species. The alternation of generations occurs in the structure of two vital forms or the independently existing haploid and diploid generation takes place in the large systematic group of higher order vascular plants. For example, in *Polypodiophyta* the beginning of a reproductive cycle (from an origination up to a maturation of spores) proceeds on the basis of a sporophyte. After that the spores leave a maternal plant, which continues further development. The spores or the rejuvenated generative cell gives rise to a prothallium or sexual generation (gametophyte). This is a more energetically strong structure as compared with a sporophyte. Therefore a prothallium represents the material substance, which is necessary for the end of the reproductive process. On a gametophyte body the development of the last and most energetically strong generative cells (ovicells, spermatozoids and zygotes) is renewed and ends. The prothallium after a zygote germination and a young plant develops perishes.

The Means of Energy Protection and Their Evolution

The questions of energy protection of plants (physical nature, principle of action, reasons for the change of mechanisms, ways of protection during evolution, etc.), which will be considered in this chapter, are new and are explained only by the wave natures of sexual gametes and the process of sexual reproduction. Unfortunately, at present in biology is the only accepted conception of organization of living matter – the corpuscular one. All the biological models were developed on the same theoretically based (cell is an open and non-equilibrium system). We have tried to expand our ideas about its corpuscular-wave nature by means of the separation and study of corpuscular and wave components in various biological processes and on the different levels of plant organization.

A development of plant order from the corpuscular-wave positions is a dual process also. It has a strict evolutionary trend, is characterized by directional alterations of a DNA-particle, DNA-wave and energy protection methods, which leads to the occurrence of a more progressive species. Naturally, all problems of the individual development of plants are probably not analyzed in the whole volume. Therefore our primary task during the next two chapters is the following. To attempt to explain the united and complex reproductive process divided in stages dependent on their functional meaning, and then in turn we will study each group of stages with corpuscular-wave positions. In this discussion we have resolved to begin on the problems of energy protection. For this argument and the description of the natural and evolutionary development of the protection system of a sporophyte, we will be using the known factual material on embryology and the systematic life of plants (Meyer, 1946; Komarnitskii et al., 1975; Poddubnaya-Arnoldi, 1976; Life of plants, 1978; Singh, 1978; Smith, 1980; Kozubov, 1982; Embryologe of angiosperms, 1984; Grant, 1984; Kozubov and Muratova, 1986; Vasiliev et al., 1988; Raven, 1990; etc.). But the interpretation of them is the other, because this phenomenon is caused by the wave properties of generative cells.

The general regulating and motivating forces of the evolution of animal and plant life, apparently, is the constant increase of the energy level of the species (by a DNA-wave line). This statement is proven by a progressive development of the energy line of a reproductive cycle, the increase of the number of stages and divisions, which participate in the process of a DNA-wave lifting to higher energy levels. The biological meaning of a protection system is the another – a regulation of interrelations between a generative and vegetative sphere of plants, a creation of optimum schemes of the life cycles of a species, which provide the viability and stability in space and time at every stage of their evolutionary development.

Once again we will remind you, that a speciation is the directed process which conducts a progressive increase of the energy potential of a species. The plant order has created the four ways of energy protection which provide the defense of a vegetative generation in the corresponding periods of development of a generative sphere. These methods have been named: haplont type, external fertilization type, bimorph type and double protection type. Accordingly, the whole evolution of the sexual process is divided on the five 5 periods dependent of the methods of protection, which a species is used for this aim (Figure 43). The given system was absent at the initial stage of the organization of the sexual process. However in the course of its development the necessity of origination of a protection system became a question of the living matters survival. In the beginning the activating mitosis or form energy transfer from the chromosome to the chromosome appeared on the basis of the former scheme of the life cycle (a stage of oogenesis in Volvox). After that the haplont type scheme of a life cycle was created. In this new matrix of a reproductive cycle the previous pattern of the transference of a DNA-wave to a higher energy level, which leads to the death of the maternal organisms (prokaryotic type) is completely absent. A process of speciation on a standard scheme of a haplont type will continue for thousands of years, until the reserve of the capacity of the protection system has been fully exhausted. The crisis of mutual relations between a generative and vegetative sphere of species was preceded at each change due to the protection methods. Then nature created a new more effective way of protection. Change of protection methods is a complex many-staged process. The cardinal change of the matrix of a reproductive cycle and the creation of the scheme of a life cycle of a new type occurs through a large number of non-viable intermediate matrices. Therefore a species, which has originated on that basis, will rapidly become extinct. Thus, a speciation is a wave process which after the establishment of a matrix of a new type of periodic expanding, and when changing the protection methods a periodic narrowing, but always directing the growth of the biological (internal) energy of a species.

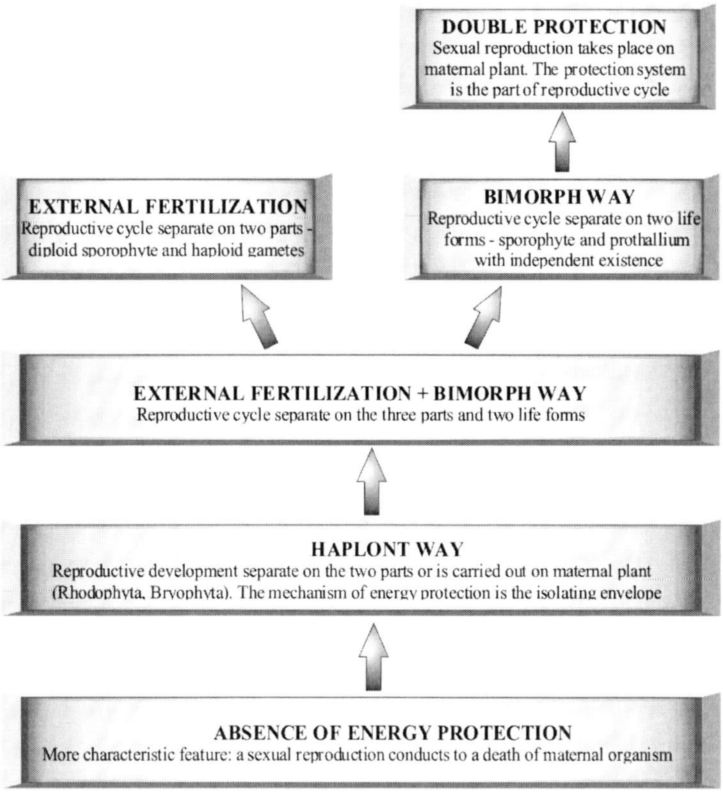

Figure 43. The scheme of the evolutionary development of energy protection ways in plants.

In figure 43 the ways of energy protection in of order of their occurrence during evolution are shown. These ways often (but not always) are connected with a specific type of the life cycle of plants – haploid, diploid and haploid-diploid. The mechanisms, which use a variety of methods for protection during the different periods of the evolutionary development of a sexual process, as rule, are the same: 1) the formation of abscission layer (with a restraining stage of development or without it) and 2) the separation of a reproductive cycle in a space. The distinctions are the following: what type of life cycle the species has ; before what stage of reproductive cycle (meiosis, fertilization) take place its rupture; that an organ of propagation (meiospores, zygospores, seeds) serve, etc. These and other differences have been altered on the basis of a division of the protection system into five periods. The more detailed description of the methods of energy protection will follow in order of development and their organization during evolution.

1) *Absence of energy protection*. This system's absence in the single-celled species *Chlamydomonas, etc.* and in the colonial forms *Gonium, Pandorina,* etc., at the beginning of a sexual reproduction leads to a maternal organisms death. The techniques of the rejuvenation of generative cells are mainly of a prokaryotic type or a prokaryotic type with activating divisions.
2) *Haplont method.* In plants with a haploid type of life cycle are found: algae (many species of Chlorophyta, Rhodophyta) and mosses (Bryophyta). The apical point of the sexual process is a meiospore or a haploid cell, which has a full program of development based on a DNA-particle and a DNA-wave. The organs of propagation are a zygospore and meiospore. The reproductive stage, before which there is a rupture of the reproductive cycle, is meiosis.
3) *External fertilization*. This corresponds to animals with external insemination and in *Cyclosporophyceae* algae. All of them have a diploid cycle of development. At the apical point of the sexual process and organ of propagation are the ovicells, anterozoids and zygotes. The stage of a reproductive cycle, which is realized outside of a maternal organism, is the fertilization.
4) *Bimorph type*. The given means of protection attempts to spread among some classes of brown algae (Isogeneratae, Heterogeneratae) and in all vascular sporous plants (Polypodiophyta, Lycopodiophyta, Psilotophyta, Equisetophyta). A life cycle is of the haploid-diploid type with the isomorphic and heteromorphic alteration of the generations. The sexless generation (sporophyte) of the majority of species is dominated. At the top of the sexual process is a zygote (*Laminaria, etc.*) or a meiospore (*Cutleriales, etc.*). The organs of propagation are the following two – a meiospore and zygote.
5) *Double protection*. The means of double protection appeared for the first time in the higher order seed plants (Gymnospermae, Angiospermae). It operates on the basis of the haploid-diploid type of life cycle with an alternation of sexual and sexless generations. A gametophyte undergoes a further reduction and has lost its own independence. All process of reproductive development of seeds proceeds on a sporophyte body. At the apical point of the sexual process is an initial cell of sporophyte. The organ of propagation is a seed.

Let us examine each of the listed methods in more detail.

1. *Absence of protection*. The system of energy protection is absent in some species of Volvocales: both in unicellular (*Dunaliella, Chlamydomonas*) and colonial (*Gonium, Pandorina, Eudorina, Volvox*) forms. All of them in a vegetative condition are characterized by the active mode of life. The sexual process is presented in different forms (hologamy, isogamy, anisogamy and oogamy). The whole body of a unicellular individual of *Chlamydomonas* and all cells of *Gonium, etc.* coenobium takes a part in the formation of gametes. The biophysical techniques, by means of which a DNA-wave reestablishes its energy state, is due to the prokaryotic type, when a cell division take place in the fixed

volume and directs the vegetative generation to its death. It is the most primitive technique for a rejuvenation of generative cells. It is used by the prokaryotic cells for the transfer of a DNA-wave with a 0th energy level to the 1st energy level during its cell divisions. It is necessary to notice, that the more progressive techniques - the activating mitosis and activating meiosis - already appeared in the highly developed systematic groups of Volvocales.

This group of algae always has a cell envelope above the plasmatic membrane. A vegetative haploid cell after entry into a process of sexual reproduction becomes a gametangium. The growth of the generative cells in fixed volume of gametangium is limited. It directs the increase of intracellular pressure and the diminishing of gamete size in a series of cell generations. The number of cell divisions is fixed as well - five or six depending on the durability of the envelope. The response of gametes due to the jump in pressure becomes the passage of the DNA-wave to the higher levels of energy. Therefore, a genome of generative cells is rejuvenated after every cell division. Rejuvenated isogametes or micro- and megagametes leave the former maternal cell after the destruction of its envelope. All gametes have the identical sex, DNA-particle and DNA-wave and the energy potential, which is, naturally, higher in the case of the formation of 64 gametes. In an external environment they are meeting with the gametes from the different male and female gametangiums. The sex gametes are attracted only to their opposite, even if they are the same by external signs. They are fused and produce a diploid zygote. The zygotes after a dormant period and a reducing division initiate the development of next vegetative generation. The meiosis includes the mechanism of the separation gametes by sex as a consequence a common pool of individuals consists of 50 % female and 50 % male genotypes. Then the period of vegetation is realized, which occurs in a haploid condition and ends by a period of sexual reproduction. Hence, the cycles of the development of vegetative and generative generations do not coincide. Therefore there is no need for a system of energy protection at the initial stage of the development of the sexual process.

The given evolutionary branch, despite a high level of the organization of a sexual process in *Volvox* and differentiations of their vegetative part, has completely exhausted the resource for the continuation of the speciation process. The growth of an energy potential of species demanded the cardinal reorganization of the matrix of a reproductive cycle. During evolution such a matrix, in which gradually disappears the prokaryotic type of techniques for a rejuvenation of generative cells, and the first method of energy protection on the basis of a haplont cycle of the development of plants has been created.

2. *Haplont method.* For the first time the haplont method of energy protection occurs in green algae (*Conjugatophyceae*). This method of protection occurs in many species of *Chlorophyta*, *Rhodophyta* and in all species of *Bryophyta*. The passage of a sexual reproduction to the more progressive technique of the "recharge" of a biological harmonic oscillator by means of the transference of biological energy from a chromosome to a chromosome during special cell divisions (activating mitosis and activating meiosis) preceded its origination. The given technique can occur in one or in a few generative cells and not direct the vegetative generation to destruction. The safety of a maternal plant with the haplont type of protection during joint development with the energetically strong zygote guarantees a strongly thickening and durable external envelope. The meiosis and formation of a meiospore, which is a more energetically strong generative cell of the reproductive cycle, is created from a maternal plant.

In the beginning the matrix of the new model of a reproductive cycle still includes the prokaryotic technique of the rejuvenation of generative cells (*Conjugatophyceae*). The most known among *Zygnematales* is the algae of *Spirogyra*. Its thread consists of cylindrical cells with a cellulose envelope and a slimy covering on the outside. A generative cell can become any cell of algae. The sexual process proceeds by a type of conjugation. The male and female gametes are considered the cells from two different threads of *Spirogyra*, which are the same by the morphological signs. However, the attraction will only be between those threads, which pertain to the different sexes of the plants (see Chapter 5). In this case the forces of attraction are so great, that the envelopes between the two cells are destroyed and the contents of the cells of one thread pass in an opposite cells of other thread. It causes the jump of intracellular pressure inside of the conjugated cell. A fusion of the genomes of male and female gametes and a passage of the double set of chromosome-waves on the higher levels of energy occurs by means of high intracellular pressure. It is a prokaryotic type of genome rejuvenation. Then the energetically strong diploid zygotes are covered by a thick three-layer envelope, pass into a rest state and remain on a body of algae. In the autumn they fall on the basin bottom due to the dying off of the threads of *Spirogyra*. In the spring the meiospores, which once again are rejuvenated during the activating meiosis (i.e., a dying off of three of the four gametes during a reducing division), germinate and give rise to the haploid individuals of the next generation, which have separated on the male and female genotypes during meiosis, despite a lack of external distinctions between them. Thus, the main part of the reproductive development of Spirogyra (from conjugation and up to zygote formation) passes on a maternal plant. The energetically strong zygote can remain a long time on the body of a maternal plant, not doing any harm. The external envelope has a sufficient thickness, to guarantee the energy protection and viability of vegetative and generative generations during a period of their joint development.

The sexual process of plants on the basis of a haplont method of protection has passed through a long evolutionary development. The nature of it during this period has created a great number of variations in a life form, some of which are unicellular, noncellular (siphonogamic), colonial, filamentous and relatively high-differentiated forms. Plants have moved on to a sedentary life. A wide spectrum of variations of the sexual process (hologamy, isogamy, anisogamy and oogamy), the techniques for a rejuvenation of generative cells (prokaryotic type, transference of biological energy from a chromosome to a chromosome), the mechanisms of energy protection, the male and female gametes with a different morphology and behavior are created. Well documented, is the principle of operation of the haplont method in the evolutionary dynamics that allows us to understand the nature and significance of a energy protection system in a life of this systematic group of plants. The initial link of the new evolutionary branch is *Ulotrichophyceae*, the sexual process of which shows how the haplont method of protection was created.

The sexual reproduction of *Coleochaete* is the example, when the envelope of zygote does not completely execute a protection function. The main part of its reproductive cycle – from an origination of oogonium and antheridium and up to a zygote formation, proceeds on the body of algae. The meiosis is carried out after a long dormant period and out of a maternal organism, like in *Spirogyra*. The formation around a zygote of an additional envelope is the characteristic feature of the sexual process of a *Coleochaete* species. This is a layer of cells, which covers it from the outside like a bark. The development of a bark is possible to explain if we proceed from the wave nature of this process as a partial failure of a protection system.

We will admit, the biofield of a fertilized ovicell is not completely extinguished by an envelope of zygote, and spreads out of its limits. In this case, the energetically weak cells, which are situated closely around the zygote and in the zone of the action of the biofield forces, begin to move. They form an additional isolating layer around the zygote. This layer remains after that, as its cells die off and the zygote passes into a resting state. The double envelope will protect the *Coleochaete* as long as the zygospore remains on the body of algae. A formation of mitospores (zoospore) at vegetative propagation becomes a frequent cause of the death of a maternal plant. Each cell of *Coleochaete* in a favorable environment can become mitosporangium and gives a beginning to 2-16 mitospores. The maternal organism in this case ceases to exist, like in Volvocales. Once again underline, that during sexual reproduction the system of energy protection operates already, which allows a resting zygospore for some time to remain on a body of algae.

The division of *Rhodophyta* applies to the most developed systematic group of algae. The sexual process of all species of *Rhodophyta* is carried out by means of the oogamy only, and has a number of essential peculiarities which distinguish them from other algae. The reproductive cycle begins from the formation of spermatogonium and carpogonium and up to an exit of the carpospores from a carposporangium is realized in a gametophyte. The zygote without a dormant period passes into the cell divisions (mitotic and meiotic), giving rise to carpospores (*Bangiophyceae*) or carposporophyte (*Florideophyceae*), which acts like a parasite on a body of algae. The apical point of the reproductive process and the organ of a propagation reproduction body is a haploid carpospore. Therefore, all species of red algae have the haplont type of energy protection, because the process of sexual reproduction is completely realized in the body of maternal plant.

In *Rhodophyta* the signs of the failure of the protection system have appeared and increased. The given failure is caused by a growth of the energy potential of a zygote, which remains on a body of a maternal plant after fertilization.

We will try to explain such a complicated cycle of development and the failure of the protection system, beginning from the fertilization, when the diploid zygote is formed. As was mentioned above the zygotes of other species of algae with a haplont method of protection after fertilization pass into the dormant state and the zygospores leave the maternal plant. However the energy potential of red algae, apparently, has reached such a high level, that the envelope of the fertilized ovicell, as the abscission layer, is not as capable of translating the energetically strong genome of zygote into a condition of rest and constrains its development by the frameworks of the one cell. As a consequence, the zygote germinates and then passes through the restriction method of development directly on the body of the maternal plant. This phenomenon morphologically will be shown in the form of threads, which grow inside a gametophyte. The restriction method of development ends by a formation of a special kind of sporangium (carposporangium), reducing the cell divisions and maturation of carpospores or haploid meiospores.

However the greatest crisis in the haplont method of energy protection is observed in mosses. The *Bryophyte* among a species with the haploid cycle of development, certainly, are the most evolutionary advanced ones. They have a high enough level of organization of the reproductive cycle. The structure of the archgonium is similar with *Polypodiophyta* and *Gymnospermae*. Therefore, all of them often unite in the one group named archegoniate. The apical point of the reproductive cycle of mosses is the meiospore. The whole cycle of reproductive development, like in the red algae, continues on the body of maternal plant (gametophyte). If we proceed from the complexity of their reproductive cycle, it is caused by the energy potential of the zygote and became still as above. The thickness of its envelope, as the abscission layer, is insufficient for a transition of a zygote into a rest state. Therefore it germinates and begins to form the diploid generation on the body of a gametophyte. However, unlike the red algae, the zygote develops a sporophyte on the outside of a maternal plant. The area of contact between the energetically strong diploid tissue of a sporophyte (sporogon) and the energetically weak tissue of a gametophyte in this case is limited. A sporogon is connected with the tissue of a gametophyte by means of the foundation, which serves for the sporophyte as a substrate and a source for a supply. At the boundary between the energetically strong tissue and energetically weak tissue begins to form the abscission layer. This barrier gradually restricts and then stops the development of an energetically strong diploid generation, preventing, thus, the gametophyte from complete exhaustion and death. However the sporogon during this restriction method has enough time for the completion of the reproductive process, namely, the creation of the necessary reproductive organs, realization of a reducing division, origination and maturation of the meiospores.

During the reclamation of the land the plants are proven to be at a crossing of three roads (Figure 44). The previous haplont method is no longer able to perform the protection function, replacement and the nature of its search methods down the new paths of its own development. The reproductive cycle of brown algae, apparently, had the more suitable matrix for the change, test and selection is the most reliable means of energy protection among them at this stage of the development of the plant world.

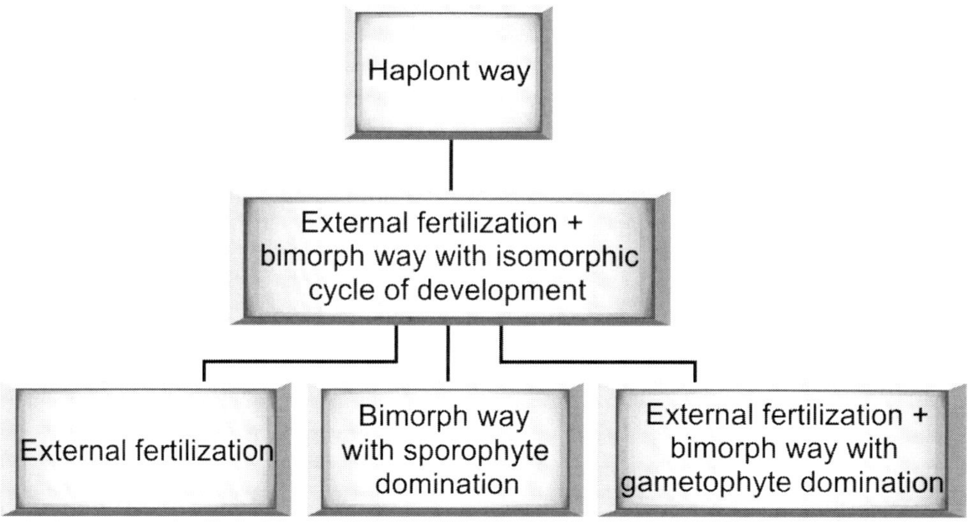

Figure 44. The evolutionary development of the methods of an energy protection system in *Phaeophyta.*

From the aforementioned it becomes obvious, because of the haplont method of energy protection in *Bryophyte* has completely exhausted its own resources, and the speciation could not proceed on this basis any more. After failure of a protection system arises a question on necessity of its replacement by one of the alternative methods, which by this time are already were created in a brown algae (*Phaeophyta*). There are: 1) external fertilization from the favorable position of a diploid generation and 2-3) bimorph types + external fertilization from a favorable position, accordingly, sporophyte and gametophyte, 4) bimorph type, which will be considered below (Figure 44). Ultimately, the new means of energy protection was found and the process of speciation was renewed on the basis of a bimorph type with a domination of a diploid generation in ferns (*Polypodiophyta*). Changes in the means of protection always directs to a cardinal reorganization of a reproductive cycle and organization of the plant life cycle as a whole and at the same time gives a powerful impetus for further evolutionary development.

2. *Bimorph way + external fertilization.* For the first time this new protection type has appeared in the brown algae. In the majority of the species of *Phaeophyta* it operates in conjunction with two methods – the bimorph method and external fertilization. The separation of the reproductive cycle on three independent parts and the alteration of generations are characteristic of its features. The process of reproductive development is carried out on the body of: 1) a haploid gametophyte, 2) diploid sporophyte and 3) during external fertilization, when the spermatozoids and ovicells are formed into an antheridium and oogonium and after that leave the environment, where their fertilization occurs. The sexual process may be of an isogamy, anisogamy or oogamy type. The life cycle of brown algae development organized as an alternating of the two generations (haploid and diploid), which live a life of independence and form the two different biomorphs.

The evolution of plant order represents an uneven process by the improvement of relations between a vegetative and generative sphere of plants. The evolutionary trend of the system of energy protection development is directed based on the use of optimal mechanisms and schemes of life cycles, which provide a realization of the whole process of sexual reproduction on a body of a maternal plant. It is necessary to underline, that the placement of the whole reproductive cycle on a vegetative generation has been reached by the creation during evolution of the first haplont method of energy protection, namely in red algae and mosses. However, the increase of disbalance between the generative and vegetative sphere demands a cardinal reorganization of all life cycles of plants. As consequence the reproductive cycle "broke up" to the three parts, and the reproductive development began to be carried out on the basis of two different vital forms (biomorph). In the evolutionary plan, the division of a reproductive cycle and the organization of a life cycle as the alternation of the two independently existing biomorphs was the step back. However, the crisis of relations went too far, and in the brown algae, instead of a haplont method, the two new ways of protection originate – a bimorph method and external fertilization.

A gametophyte and sporophyte by the morphological characteristics are similar in the majority of *Phaeozoosporophyceae* (Ectocarpales, Sphacelaria, etc.). The isomorphism is a more ancient organization of a life cycle as compared with heteromorphism. The evolutionary development of brown algae proceeded in two directions: with the prevalence of a sexless generation (sporophyte) or a sexual generation (gametophyte). Therefore the whole cycle of the individual development of the species with the isomorphic type represents the alternation of two morphologically identical by outward appearance of haploid and diploid vital forms which lead to their independent existence. After the divergences of the species of *Phaeozoosporophyceae* in one part of its representatives (*Cutleriales*) the sexual generation prevails, and the sexless generation is the small plate, which is pressed into a substratum. In other species (*Laminaria*) the sexless generation, on the contrary, dominates, and the sexual generation is presented by a small non-differentiated haploid prothallium (male and female), which are formed after a germination of zoospores. The heteromorphic cycle with the prevalence of a sexual generation has not accomplished further development during evolution. The evolution of plant order proceeded in a direction of the dominance of the sporophyte or a sexless generation with a diploid level of an organization genome. In turn methods the system of energy protection, which was to organized in *Phaeozoosporophyceae* as the connection of two methods - a bimorph type and external fertilization, have divided on the two evolutionary branches: the class of *Cyclosporophyceae* (external fertilization) and the division of *Polypodiophyta* (bimorph type).

3. *External fertilization*. The external fertilization as the method of energy protection has gained the greatest distribution in animals. In plants it meets in the species of the one class of brown algae (*Cyclosporeae*). The originality of the reproductive cycle development of the *Cyclosporeae* lies in the fact, that it is similar to the animal cycle. I.e., it is inherent to the algae with the diploid type of life cycle organization, when only gametes are haploids and the fertilization happens outside of a maternal organism. Its typical representative is the perennial algae *Fucus vesiculosses* L. A powerful sporophyte reaches a length of 15-150 centimeters and has the dichotomous branched out thallus, which grows by means of the one apical cell. The most part of reproductive cycle is realized on the sporophyte, and fertilization – outside of. The formation of zygote is the final part of reproductive development of algae. In an antheridium are formed the 64 anterozoids, which after maturing are pushed out from the scapidia into the water. The oogonium develops on the ends of the thallus and into the special deepening scapidia. After the reducing division the 8 ovicells are formed in the oogonium. The meeting and fusion of male and female gametes occurs outside of a maternal plant and without its participation. An energetically strong zygote, which is also the organ of propagation, without a rest period germinates in a new plant of *Fucus vesiculosses*.

The success of sexual reproduction, the protection of which is organized by means of external fertilization, strictly depends on the development degree of the operational functions of organisms (a search of partner). In animals, which lead a movable life and have different devices for the external insemination, it ensures reproductive success and formation of a viable generation both in the water and, especially, on the land. For plants with motionless life, the given method is not appropriate, especially after exit on the land. Therefore another method (bimorph) of energy protection is widespread in plants and replaced the previous haplont method. This method is characterized by the alternation of generations, which lead an independent existence.

The special place for the understanding of the nature of the mechanisms of the separating of energetically strong generative cells from the maternal organism belongs to the *Diatomeae*, in spite of that this systematic group of algae originated later than the brown algae. During their path of evolution it is possible to track all stages of separation of a zygote (auxospore) from the initial diatomic algae. After fertilization the rejuvenated zygote begins to realize the potentialities received during activating meiosis for development by means of rapid cell growth and restores its maximum size for each species. The auxospores of *Centrophyceae* are formed on the place of a maternal cell. In a *Pennatophyceae* they participate in fertilization after a rapprochement of two individuals, and the generative cells, which coming out from them, are fused into a single zygote. In the last case an auxospores in dependence of the position and in communication with a maternal organism are differed, connected (lateral, terminal, etc.) and free zygotes.

Once again we will remind you, that all species with external type of fertilization are the diploid organisms. A life cycle of diatomic algae is a good model for studying the nature of meiosis and fertilization too. The process of the decrease in sizes during vegetative propagation which can be stretched for two or more years directs the growth of intracellular pressure (see Chapter 3). The mechanism of activating meiosis put in action by means of pressure forces, which ends by a formation of four hereditary and energetically different generative cells. As a rule, one or two energetically strong meiospores are formed. The energetically strong gametes participate in fertilization, and the energetically weak nuclei or cells perish. All diatomic algae are diploid organisms. The period of vegetative development passes in a diploid condition, and a reducing division precedes the fertilization. Therefore the given systematic group of algae can be considered as an important step in the evolutionary development of plants, during which time there was a transition of plants on the diploid level of life organization. The confirmation of this hypothesis is the phenomenon of autogamy in some species of *Centrophyceae*.

Let us analyze how to change the state of the genome of generative cells, which in the course of activating meiosis forms two viable and two nonviable gametes. Also it is necessary to consider that the two of them are the male sex and the two gametes are the female sex. The maternal cell is separated from the environment by a dense flint armor wall and the external noises will be on a minimal level. Therefore a biological harmonic oscillator of a diploid maternal cell in regards to the four oscillators of the haploid gametes serves as the synchronizer for their oscillation. As a consequence, the frequency and phase of oscillations of a DNA-wave in the four gametes (energetically strong and weak, male and female) is identical. In the process of fertilization the energetically strong male and energetically strong female gametes take part. During syngamy they are attracted as the two complex objects with opposite spins and formed is a new stable genetic structure – a nucleus with the diploid set of chromosomes and the haploid set of resulting waves with greater energy, because the amplitudes of oscillation add up. The peculiarity of the spatial disposition of chromosomes into an interphase nucleus is the close arrangement of homologues (Monachova, 1990). The homologous chromosomes are attracted to each other, because its chromosome-particles are identical in size, form and structure, and the chromosomes-waves coincide by the frequency and phase of the oscillations. Therefore the formula of the genome of a diploid cell represents 2 DNA-particle and 1 DNA-wave. A rejuvenated zygote then enters the next cycle of development. The two cells of diatomic algae with the completely discharged biological harmonic oscillators at this time are destroyed and die.

4. Bimorph type. The use of external fertilization as the way of energy protection is a temporary measure, because for terrestrial plants it is not suitable. The external fertilization disappears from the scheme of the reproductive cycle after the plants exit from the water onto land, where the meeting of gametes in this way is more difficult. Only the one way of energy protection - the bimorph type passes at the evolutionary process. It will execute the defensive function in *Polypodiophyta* until the emergence of seed plants. The essence of the bimorph type is that the first stages of the reproductive cycle (up to the maturation of spores) occur on the sporophyte. Then the spores leave a mother plant. In favorable conditions they germinate and give rise to a new morphological structure - prothallium. The reproductive process is renewed and the last stages of the generative cells development (oogenesis, fertilization, etc.) take place on its body. A gametophyte is a rejuvenation tissue (because an energetically strong spore gives rise the gametophyte) as compared with the sporophyte. But even it was not able to translate the zygote (the most energetically strong generative cell of reproductive cycle) in the resting state. That is why the zygote after fertilization germinates on a gametophyte and begins to form a young fern plant. At the beginning it grows by means of the nutrients from the prothallium, which soon dies. As regards to a sporophyte, it continues its ontogeny at this time.

A link between the haplont and bimorph methods of energy protection is clearly seen. But the degree of relationship and the precise origin of *Polypodiophyta* are hard to establish, because a great number of intermediate species in the overall chain of organisms has disappeared, and this transition simultaneously has touched a number of vital systems, associated with the cycle of ontogeny and reproduction. These are: 1) a separating of the energetically strong zygote from a gametophyte and the origination of the two independently existing vital forms - a sporophyte and a gametophyte; 2) a change of the motion direction of biological energy during the formation of the energy program of the future plant, and 3) a more powerful development of the sporophyte and in parallel with that, a reduction of the gametophyte. The generative sphere of brown algae is on the lower level of development. Therefore, most likely, the ancestors of *Polypodiophyta* were the highly developed species of *Bryophyte*. This is the conclusion we have come to during the corpuscular-wave decoding of their reproductive cycles. The reproductive cycles of mosses and ferns, such as the primitive class of *Ophioglossidae*, consist of the identical number of stages for a change of a DNA-particle and a DNA-wave, but they have another arrangement concerning each other. It is likely that the transition to a *Polypodiophyta* happened as follows.

As mentioned the zygote is the most energetically strong generative cell, which is formed on the maternal plant, in the group of algae with the haplont type of protection. At the end of reproductive development the zygote passes into the rest state and completely separates from the body of the plant by means of a thick cell wall or abscission layer. The zygote of red algae germinates on the maternal plant and occurs in the restriction stage of development, which ends by the formation of carpospores. In mosses it gives rise the sporogon or a diploid generation, a formation of which takes place outside of plants. The next step, which proceeds from the trend of evolutionary changes, will follow. The zygotes after fertilization should have left the maternal plant and began independent living, like the sporophyte. The mechanism of separation is well documented by the example of diatoms (as was discussed above). Thus, the haplont type disappears and turns into the bimorph method of protection.

A reproductive cycle of mosses and ferns is organized in such a manner that the sexual cells develop both on a haploid tissue of gametophyte and on a diploid tissue of sporophyte. However the "bioassemblage" of the energy program in the mosses begins on a base of a gametophyte, and its continuation happens on a base of a sporophyte. i.e., the energy vector of increased biological energy is directed from the gametophyte to the sporophyte. The apical point of the reproductive process is a haploid spore. The given process in ferns goes to an opposite side – from the sporophyte to the gametophyte with the maximum level of biological energy in a zygote. During evolution the capacity of the energy line is increased due to two parts - on the basis of gametophyte and sporophyte, but the rates of growth of a biological energy were higher in the reproductive structures, which develop on the gametophyte. For example, in mosses the archegonium or the new female reproductive organ, whose functioning is connected with a DNA-wave transition on the higher energy levels, has appeared. Let us assume that during any moment in time the power capacity of an energy line situated on the gametophyte moved above the line of sporophyte. In this case the direction of stream of the biology energy must be changed to the contrary - from the sporophyte to the gametophyte, when the reproductive development is initiated by a diploid sporophyte, and it ends on a haploid gametophyte by the zygote formation. Hence it follows that the matrix of the reproductive cycle of a new typical basis, which has all species of Polypodiophyta, could have been created by means of these transformations. The cardinal transformation of the reproductive cycle has given a powerful push for the development of a sporophyte, because the zygote with the diploid genome has more potentialities for development, than a haploid spore. A gametophyte, which now is presented as the prothallium, on the contrary, has started to be reduced. This change of the dominants subsequently remains in the more developed seed plants. It is necessary to underline, that given transformation is had the last and most cardinal reorganization of the matrix of the reproductive cycle, system of energy protection and the whole life cycle of plants.

5. Double protection. During evolution of higher plants originated the 5th and for today the most advanced way of protection, which is organized by the principle of a double protection. As already mentioned, one of the major trends of organic world development is the implementation of a development of a generative generation on the body of a sexless generation. This aim was achieved in the red algae and mosses. However, the process of speciation on the basis of a reproductive cycle with a haplont method of protection is deadlocked, that is manifested at first in the red algae, but most strongly in mosses (*Briophyta*). The crisis of the relationships between the generative and vegetative sphere led to the division of a matrix of reproduction cycle on a basis of three parts and then on two parts. This tendency was realized on the next stage of the evolutionary development of plants – in gymnosperms. For this purpose the new method of energy protection was created, where a sporophyte and an embryo are separated by a double "cover." The method of dual protection presents a higher level of organization in a protection system, because it has the mechanism for the increase of protective function, which is embedded into the reproductive cycle of every species.

For the first time the method of double protection has appeared in *Gymnospermae* in response to the growth of an energy potential of a species on the bases of the *Polypodiophyta* matrix of reproductive cycle. The most distinctive trait is, first, the presence of the alternation of generations with strongly reduced gametophyte, which has lost the ability of an independent existence. Second, the process of sexual reproduction is completely carried out on a sporophyte. In higher order plants the formation of the main structural elements of a future seed (a zygote, functional megaspore and central cell) are included in the matrix of the reproductive cycle. In angiosperms double fertilization has appeared. The apical point of the reproductive process is the initial of a sporophyte (a tier of embryo cells of 16-cellular embryos in the gymnosperms and an apical cell of 2-cellular embryos in the angiosperms). The initial cell, as the last and most energetically strong generative cell, begins to divide on a sporophyte. Then it passes through a restricting stage of development (or period of embryogenesis) in the tissue of endosperm, passes in a condition of rest and becomes an embryo. Thus, the seed, or the new structure of sexual reproduction and organ of propagation, develop. It consists of a peel, an embryo and a nutritious tissue of endosperm.

In seed-bearing plants the seed formation completely occurs on the body of sporophyte. It is the strictly determined reproductive process. The regulation of relations between a generative and somatic tissue we consider on the example of the generative sphere of *Pinus sylvestris* L. It is known, that the beginning of sexual reproduction is connected with activity identified as floral meristem. A meristem is the most metabolically active (i.e., energetically strong) somatic tissue of a maternal plant. The reproductive cycle of pine consists of the two lines, on which is formed the hereditary (on the basis of DNA-particle) and energy (on the basis of DNA-wave) programs of future plant development. This question in more detail will be analyzed in the following chapter. Here we will touch upon the questions, which are connected only with the action of a protection system.

The energy line of the reproductive cycle of the *Pinus* species includes the six stages by the increasing of biological energy (Kuznetsova, 1998, 2002). A functional megaspore is formed on the second of such stages - a megasporogenesis. During activating meiosis (or female meiosis) a DNA-wave passes to the higher energy levels, and the megaspore becomes the energetically strong cell. The energy gradient appears between it and the energetically weak cells of nucellus. A megaspore begins to divide, passes the restricting stage of development during megagametophytogenesis and gives rise to the female gametophyte. The tissues of the endosperm and a nucellus come into the equilibrium state by means of a special abscission layer - the sporoderm. The third stage caused by the rising of biological energy is oogenesis. An archegonium initial cell (one of the cells of female gametophyte) undergoes the two divisions of activating mitosis and increases its energy potential. As a consequence an energy gradient originates between the central cell and the near-by cells of the endosperm. The central cell is quickly increased in size and is formed during maturation of the ovicell, which comes in the equilibrium with the endosperm by means of the somewhat thickened ovicell envelope (Figure 41-c). Thus, a double equilibrium is achieved on the sporophyte before the fertilization - between the endosperm and the nucellus and between the endosperm and the ovicell. The fertilization breaks this balance and the female gametophyte goes out from the resting state. The ovicell and sperm have participated in the zygote formation, while the endosperm provides a protective and trophic function in relation to a future embryo.

The proembryogenesis is the last reproductive stage, which is associated with the transition of a DNA-wave to higher energy levels. A zygote undergoes the two free-nucleus divisions and the two cell divisions (activating mitosis). After that a reproductive development of initial sporophyte (or the tier of embryonic cells of a secondary proembryo (Kuznetsova, 1998, 2002)) is terminated and its ontogenesis period is beginning. Before ontogenesis the initial sporophyte is transferred from the ovicell to the tissue of endosperm. The energy gradient appears between the embryonic cells and the near-by cells of endosperm. An initial sporophyte, as the energetically strong structure, begins to divide and forms the embryo inside of the endosperm. Its cells on a common boundary are destroyed and the nutrients are absorbed by the rapidly growing embryo. During embryogenesis between an embryo and endosperm the abscission layer (embryonic chamber) is formed. This layer separates the embryo from the nutrients of endosperm and translates it into a resting state. Therefore the embryogenesis presents the restriction stage of the development of the initial sporophyte, which is carried into the endosperm or the tissue with more stronger energy potential than any somatic tissue of a plant. During embryogenesis between a generative and vegetative sphere has reached a new equilibrium, when the embryo has separated from a sporophyte by means of the protecting tissue (endosperm) and the two isolating layers - the embryonic chamber and seed peel. Thus, the regulatory role of the endosperm in these processes becomes an obvious fact. The important evolutionary achievement has been the inclusion of the protection system in the matrix of the reproductive cycle. As a consequence the growth of the energy potential of species directs the automatic increase of the protective function of the endosperm. In gymnosperm the formation of endosperm take a part in the first two stages of a female generative sphere - a megasporangiogenesis and megasporogenesis.

In angiosperms the way of double protection has been improved and has risen to a higher level of organization. The main principle of its action in the angiosperms and gymnosperms has not changed. First, the energy potency and, accordingly, the insulating properties of the endosperm tissue are planned out and regulated by the matrix of the reproductive cycle, which has specific guidelines for each species structure. Second, the energy gradients for each species in the two systems - sporophyte-endosperm and endosperm-embryo - are genetically determined. However in gymnosperms in the formation of the energy potency of endosperm (functional megaspores) involves only two stages of the reproductive cycle. In the angiosperm for a given process all stages of the reproductive cycle up to the fertilization are involved and such phenomenon occur such as the double fertilization, which gives rise to the triploid endosperm, is originated. The protection potency of a triploid endosperm is created by the fusion of 3 generative nuclei. A double fertilization significantly increased the protective properties of the endosperm tissue.

The protection system in seed plants resembles "the nesting doll". All generative organs are laid inside the meristematic tissue, or the most energetically strong tissue of plant. Further, by the course of seed formation, in succession "are inserted" into each other in the following order: a seed coat, endosperm, embryo chamber, and embryo in the center. The embryo is in a rest state. If the seeds for the some reason are left on the maternal plant, then they themselves are not able to leave the resting state and are doomed to death. The seeds for germination should leave it and come to be in favorable environments.

Conclusion

During evolution a plant order has created several types of matrices and a great number of species-specific reproductive cycles on their basis. The four universal mechanisms enter into the structure of the reproductive cycles of all plant and animal species, which reproduce sexually. These are the cell divisions connected with: 1) the change of a DNA-particle; 2) the transition of a DNA-wave to the higher energy levels; 3) the differentiation of gametes by sex, and 4) the multiplication of the number of reproductive structures. The main driving force, which directs the improvement, development and independence of the species from the environment, is the progressive increase of their energy potential on the basis of a DNA-wave. The rigid structure of the matrix of the reproductive cycle ensures the continuity of generations during changing of the haploid and diploid stages of development, during recharge of a biological harmonic oscillator in a reproductive initial cell and on the new cycle of ontogenesis. The initial cell for sexual reproduction is the genome (a DNA-particle and DNA-wave) of cells of the whole organism (*Chlamydomonas, Pandorina*, etc.), or the genome of only a definite part of the somatic cells of plant, which become the generative cells. In the first case, the transition to reproductive development leads to the death of the maternal organism, in the second case the whole reproductive cycle or its parts is carried out on the body of maternal plant. If the generative cell development takes place on the mother plant, then the relationships between the vegetative and generative sphere in each taxonomic group adjusted and are described in this chapter as the different ways of the organizing of a protection system.

The evolutionary changes were not clear until we realized that the nature of the protection system should be conditioned by the oscillatory properties of the two interacting systems - generative and vegetative. When a DNA-wave increases its own energy level, then the gradient between a rejuvenated reproductive structure and the surrounding tissue originates and begins to act the processes, caused the wave nature of them genome (regime of capture, centripetal motion of biomolecules, etc.). The biofield of energetically strong cells attracts and forms to neighboring cells in a radial rows, transfers them on the tall mode of operation, creates a centripetal stream of metabolites, as a consequence of that the reproductive structures are increased in volume. The development of them is possible to stop only by means of the removing of the last generative cells from the body of the maternal plant or pass them into a rest state (with or without a restriction stage of development). All revealed ways of protection are founded on the use of only these two mechanisms and its combinations.

The investigation of phylogeny and reconstruction of the sexual process has confirmed the important role of the system of energy protection in the evolution of plants. The system of sexual reproduction has passed the several stages of evolutionary development from a lack of protection to the formation 4 methods: haplont type, external fertilization, bimorph type, double protection and created the 5 schemes of energy protections on their base in the following order: a haplont type → an external fertilization → a bimorph type + external fertilization → a bimorph type → a double protection. The haplont type has the species of plants with the haploid cycle of development. The crisis of this way of protection has literally "torn" the reproductive cycles of brown algae into three parts and has given rise to the two new methods - an external fertilization and a bimorph type, which initially acted together. Then these ways parted: the external fertilization has received a wide distribution in animals and the bimorph way came to dominate the *Polypodiophyta*. A long path of evolutionary development is needed in fern plants; before the integrity of the reproductive cycle was restored it was created by way of double protection in the gymnosperms. In seed plants the reproductive development was carried out on a sporophyte up to the seeds maturation. The way of double protection is very effective and reliable, because the mechanism of formation of the protective properties was included into the matrix of the reproductive cycle. Accordingly, during speciation the energy gradient between the initial cell of an endosperm and the initial cell of an embryo is regulated by the structure of its own reproductive cycle and is determined for every species by specific values. During evolution the protective properties of endosperm or the intermediate protective layers are increased. So if in gymnosperms the energy potential of a functional megaspore is increased only on the two stages of the reproductive cycle, then in angiosperms the formation of a central cell, which forms the endosperm after fertilization, is involved in almost all of its stages.

In conclusion, it should be added that the process of speciation and the increasing of the energy potency of a species is still ongoing. At present, apparently, the matrix of a reproductive cycle of a new type is created on the basis of apomixis. Accordingly, the system of energy protection will be perfected and until then, the evolutionary development of plants will continue.

References

Belousov, L.V. (2005). *Bases of general embryology.* Moscow, Moscow State University.

Berg, L.S. (1977). *Works by the evolution theory 1922-1940.* Leningrad, Nauka.

Brachet, J. (1960). *The biochemistry of development.* Pergamon Press, London, New York, Paris, Los Angeles.

Buchhollz, J.T. (1946). Volumetric studies of seeds, endosperms end embryos in *Pinus panderosa* during embryonic differentiation. *Bot. Gas.*, 108, 232-244.

Chehregani, A.K., Tanaomi, N and Ranjbar, B. (2008). Pollen and anter development in *Onobrychis shahuensis* Bornm. (*Tabaceae*). *Int. J. Bot.,* 4, 241-244.

Clarckson, D. (1974). *Ion transport and cell structure in plants.* McGraw-Hill, London, UK.

Dickinson, H.C. (1990). Self-incompatibility in flowering plants. *Bioassays.*, 12, 155-161.

Embryologe of angiosperms. (1984). Ed. B.M. Johri. Springer-Verlag. Berlin, Heidelborg, New York, Tokye.

Epifanova, and Terskich, (1968). Periods of rest and active proliferation of cells in the life cycle.*Biol. Buttetin Rev.*, 29(4), 392-402.

Esau, K. (1977). *Anatomy of seed plants.* J. Wiley and Sons, Inc., New York, Santa Barbara, London, Sydney, Toronto.

Gilbert, S.F. (2010). *Developmental biology*, 9th ed., Sinauer Associates, Inc., Sunderland, Massachusetts.

Grant, V. (1984).Speciation in plants. Moscow, Mir.

Heslop-Harrison, J. (1978). Genetics and physiology of angiosperm incompatibility systems. *Proc. E. Soc. London.* Ser. B, 202, 73-92.

Heslop-Harrison, J. (1983). Self-incompatibility: Phenomenology and physiology. *Proc. E. Soc. London.* Ser. B, 218, 371-395.

Hughes, A.F.W. (1959). *A history of cytology.* Schuman, A. (Ed.), London.

Kapil, R.N. and Tiwari, S.C. (1978). The integumentary tapetum. *Bot. Rev.*, 44(4), 457-490.

Kazmin, S.D. (1984). *Biochemistry of the mitotic cycle of tumor cells.* Kiev, Naukova dumka.

Komarnitsii,N.A.,Kudryashov, L.V. and Uranov, A.A. (1975).*Systematics of plants.* 7th ed. Moscow, Prosveshenie.

Konar, R.N. and Moitra, A. (1980). Ultrastructure, cyto- and histochemistry of female gametophyte of Gymnospermae. *Gamete Res.*, 3, 67-97.

Korotkova, G.P. and Tokin, B.P. (1977). *Embryology and genetics. Debatable questions.* Leningrad, Nauka.

Kozubov, G.M., Trenin, V.V., Tihova, M.A. and Kondratieva, V.P.(1982). *Reproductive structures of gymnospermae (comparative description).* Leningrad, Nauka.

Kozubov,G.M and Muratova, E.N. (1986).*Moderngymnosperms(morphological-systematic review).* Leningrad, Nauka.

Kuznetsova, N.F. (1990). Ultrustructure of female gametophyte of *Pinus sylvestris.T cytology,* 32, 1168-1174.

Kuznetsova, N.F. (1998). Activation in the course of reproductive development in female generative sphere of *Pinus sylvesrtis L. T cytology,* V. 40(10). P. 849-854.

Kuznetsova, N.F. (2002). The reproductive development of initial sporophyte of *Pinus sylvesrtis L. T cytology, 44(2),* 159-165.

Lajtha, L. G. (1963). On the concept of the cell cycle.*J. Cell. Compar. Physiol.* V. 60. N. 2. P. 143-145.

Lehninger, A.L. (1985). *The principles of biochemistry*. V. 1-3. Moscow, Mir.

Life of plant. (1978). V. 1-6. I.V. Grushevitskii and S.G. Zilin eds. Moscow, Prosveshenie.

Ling, G. (1984). In search of the physical basis of life. Plenum Press, New York, London.

Mepham, R.H. and Lane, G.R. (1970). Observation on the fine structure of developing microspores of *Tradescantia bracteata. Protoplasma,*70, 1-20.

Meyer, K.I. (1946). Origin of ground vegetation. 4th ed. Moscow, Sovetskaya nauka.

Monachova, M.A. (1990). Cytogenetic aspects of spacial organization of interphase nucleus. *Uspekhi sovremennoi biologii, 10(5),* 163-179.

Pacini, E. (1997). Tapetum character states: analytical keys for tapetum types and activities. *Can. J. Bot.,* 75(9), 1448-1459.

Pandey, K.K. (1980). Evolution of incompatibility systems in plants: origin of "independent" and "complementary"control of incompatibility in angiosperms. *New Phytologist,* 84(2), 381-400.

Poddubnaya-Arnoldi, V.A. (1976). *Cytoembryology of angiosperms plant. Principles and perspectives.* Moscow, Nauka.

Raven, P., EvertR. and Aykhorn, S. (1990). Modernbotany. Moscow, Mir.

Rubin, A.B. (1999). *Biophysics*. V. 1-2. Nauka, Moscow.

Runions, C.J. and Owens, J.N. (1999). Sexual reproduction of interior spruce (*Pinaceae*). I. Pollen germination to archegonial maturation. *Int. J. Plant Sci.* 160(4), 631-640.

Singh, H. (1978). *Embryology of Gymnosperms.*Gebrüder Borntraeger, Berlin, Stuttgart.

Smith, J.M. (1980). *Evolution of sexual reproduction*. Moscow, Mir.

Sorokina, Z.A. (1978). Condition of potassium, sodium and water in cell cytoplasm. Kiev, Naukova dumka.

Swanson, C.P. and Webster, P.L. (1977). *The cell*, 4th edn. Prentice-Hall Inc., Englewood, Cliffs, Jersey.

Tokin, B.P. (1987). *General embryology*. 4th ed., Moscow, Visshaya shkola.

Trenin, V.V. (1986). *Cytoembryology of Larix*. Leningrad, Nauka.

Trenin, V.V. (1988). *Introduction in coniferous cytoembryology*. Petrozavodsk, Karelsky filial AN SSSR.

Vasiliev, A.E, Voronin, N.S. and Yelenevskii, A.G., etc. (1988). Botany.Morphology and anatomy of plants. 2nd ed. Moscow, Prosvechenie.

Villiers, T.A. (1972). Seed dormancy. *Seed biology*. pp. 219-281. Kozlovski T.T. ed. V. 2. Academic Press, New York.

Vishnyakova, M.A. (1997). Evolutionary succession of the structural mechanisms of gametophytic and sporophytic types of self-incompatibility reaction. *Bot. Rev., 82, 1-17.*

Willemse, M. T.M. and Boer-Dejev, M.J. (1981). Megasporogenesis and early megagametogenesis. *Acta Soc. Bot. Pol.,* 50(1-2), 111-120.

Zavada, M.S. and Taylor, T.N. (1986).The role of self-incompatibility and sexual selection in the gymnosperm-angiosperm transition: a hypothesis. *Amer. Nature*, 128(4), 538-550.

Zavarzin, A.A., Harazova, A.D. and Molitvin, M.N. (1992). *Cell biology: General cytology*. St. Petersburg, Izdatelstvo SpbGU.

Chapter VII

Corpuscular-Wave Model of the Reproductive Cycle of *Pinus sylvestris* L.

Abstract

The corpuscular-wave model of the reproductive cycle of *Pinus sylvestris* L. from the inception of microsporophyll and megasporophyll primordium and up to the formation of a sporophyte initial is considered. In pine, the sporophyte initial is the tier of the embryonic cells of a 16-cellular (secondary) proembryo.

Scots pine is a bisexual monoecious plant. A megastrobiles are located usually in the upper crown of a tree, and a microstrobiles its lower part. The matrix of a species reproductive cycle has the 14 stages of three types. The hereditary line has three reproductive stages (usual meiosis, activating meiosis and fertilization). The species-specific and the individual hereditary program in each cell of a sporophyte initial on a basis of a DNA-particle by means of a crossing-over and a gamete syngamy is formed. During meiosis and fertilization the polarization DNA by spin direction and the definition of gametes by sex take place. The energy line of a reproductive cycle consists of the two types of reproductive stages: 1) the 5 stages by definition of a reproductive initial and 2) the 6 activation stages. On the stages by definition of a reproductive initial lies the choice of the first generative cell, which passes to the activation stage and begins to undergo the divisions of activating mitosis, takes place. That is why all reproductive cycles of higher plants by the male and female generative line begin just exactly with the stage by definition of a reproductive initial. In Scots pine it precedes the processes of microgametophytogenesis, oogenesis and proembryogenesis. On the activation stages the transition of a DNA-wave to higher energy levels occurs. The species-specific and individual energy program in each cell of a sporophyte initial on the basis of a DNA-wave by means of activating mitosis and activating meiosis is formed. In this chapter we have tried to explain the idea, that the technological chain of a "biological assembly" of a genotype of sporophyte initial on the example of *Pinus sylvestris* L. presents itself on the basis of experimental data and the corpuscular-wave nature of a reproductive process.

Since ancient times the mystery connected with sex, sexual reproduction, loss and restoration of totipotency has attracted the attention of numerous scientists. Now the huge amount of experimental material – from the phenological, morphological investigations up to the unique experiments by transplantation of generative cells and tissues - has accumulated. The process of the development of sexual cells is in details described in the majority of living species of animals and plants today, and the extinct species by paleontological data. It is shown, that the mechanism of sexual reproduction represents a many-stage, genetically determined and conservative enough process. In it there is not a single random cell, a single extra cell division. Not by accident, the embryological signs are put in a basis of systematic and evolutionary studies of plants and animals. However the nature and the driving force of reproductive development are not known till now and continue to be a subject of numerous controversy and discussions.

From the middle of 19th century the embryology of gymnosperms as the scientific guidance has begun to develop rapidly in connection with the improvement of optical technique and technology of a creation of cytological micro preparations. The important contribution in the embryological investigations have brought the classical works of Hofmeister (1848, 1951), Strasburger (1872, 1897, etc.), Gorozchankin (1883), Belajeff (1893), etc., in which the description of reproductive processes and the structure of generative organs in a species of *Coniferopsida* is given. The first generalization of data by the embryology of gymnosperms that brought about the book by Coulter and Chemberlain (1917) is considered. Later there was a number of fundamental works by (Schnarf, 1933; Chamberlain, 1935; Meier, 1947; Singh, 1978; Kozubov et.al., 1982; Trenin, 1988; Tretyakova, 1990; Belousov, 2005; Williams, 2009; Gilbert, 2010; etc.).

According to our understanding, the sexual process has a corpuscular-wave nature therefore, the certain part of its mechanisms cannot be explained from a position of knowledge of modern (corpuscular) biology. Nevertheless, this work has begun with the decoding of the reproductive cycle of pine within the framework knowledge of corpuscular biology (Kuznetsova, 1998, 2002). We long searched for the reason of a duality of matrix. After a departure from the corpuscular-wave nature of DNA and the quantum mechanical nature of a cell we are faced with a new task - to define a place, a role and a behavior of wave components in biological processes. Six of the previous chapters have been devoted to these questions. In this chapter we come back again to the data of previous analysis (Kuznetsova, 2002) and the matrix of reproductive cycle of *Pinus sylvestris* from the new theoretical positions will be described once more. Particular attention is paid to the questions of polarity and asymmetry in the organization of reproductive structures. Because all higher plants are immobile organisms, the polarity of generative organs, generative cells and organelle distribution in plants can be tracked (Dure, 1976; Esau, 1980; Konar and Moitra, 1980; Fowler and Quatrano, 1995, etc.). It is considered, that gravity in the creation of cell polarity has the leading role (Sinnott, 1960; Esau, 1980; Stephens et al., 1995; Drubin and Nelson, 1996).

For the analysis of mechanisms of sexual reproduction we will begin with the gametogenesis, which leads to the formation of ovicells and sperms and due to highly specialized processes of four types. The pine reproductive cycle includes: 1) stages on which the hereditary program of development of the future plant (connected with a change of the chemical formula of a DNA-particle and chromosome number) is created; 2) stages on which the sex of the gametes is defined; 3) stages on which new reproductive initials arise; 4) activation stages on which the energy program of the development of a future plant (connected with a change of energy potential of a DNA-wave) is created. Let us consider each reproductive stage in more detail.

Stages of the Creation of the Hereditary Program

In all species of higher plants the hereditary program of a sporophyte initial is formed at three stages of the reproductive cycle: microsporogenesis, megasporogenesis and fertilization. The cytology of meiosis and the syngamy of gametes in coniferous plants are described in many literary sources (Ferguson, 1904; Owens and Molder, 1971; Singh, 1978; Kozubov et al., 1982; Trenin, 1988; etc.). It is shown, that during meiosis the chemical formula of a DNA molecule is changed by means of the special molecular mechanism – a crossing-over. During fertilization the numerous new combinations of genes are created also. In chapter 3 the hypothesis of the physical nature of the crossing-over, meiotic divisions (usual and activating meiosis) and fertilization were already discussed. On the scheme of the pine reproductive cycle we will designate the given type of stages by the symbol → ☓.

Stages by Definition of the Gamete Sex

The meiosis from a position of physics and biology is the unique cell division, the nature and significance of which in a process of sexual reproduction up to their end are still not studied. By means of meiosis, a DNA-particle change, a reduction of chromosome number, a generation alternation (haploid and diploidic phases of development) and a differentiation of sexual gametes by sex has occurred. Perhaps, the definition of gametes by sex has a quantum-mechanical nature and is caused by the next mechanism. The nucleotides into a non-uniform magnetic field, which is created by two centrioles during the leptotene-pachytene stage of meiosis, are divided on two beams in depend of the direction of their spin (see Chapter 6). As a result of nucleotide filtration a chromosome begins being polarized and the two types of sexual gametes have appeared: 50 % of the gametes with spin-up ↑ and 50 % of the gametes with spin-down ↓. On the scheme of a reproductive cycle we will designate the given type of stages by the symbols →⚥.

The phenomenon of the differentiation of generative cells by the sex had an effect on the structure of the reproductive cycle of all species with sexual reproduction. A general scheme of a reproductive cycle by this form resembles the fork (Figure 45). It is created from three main parts: the male generative line (♂), the female generative line (♀) and fertilization – a place of two lines of unification.

During evolution the sexual system passed an independent and many-staged path of development. The development of the gametic dimorphism went in the direction of the intensification of the differentiation and specialization of sexual gametes – from isogamy to anisogamy, and then from anisogamy to oogamy. Proceeding from the law of linear momentum conservation, the male gametes have become smaller and mobile and the female gametes, larger and motionless. In higher order plants the behavior and morphology of male and female gametes (size, mobility and internal organization) link with the energy state of a cells of the floral meristem of two types – a male and a female, which begin the process of reproductive development. Later on that basis the male (♂) and female (♀) generative lines of a reproductive cycle are formed. ↑). Because the megastrobiles are located in the top part of the tree crown, and the microstrobiles is in its lower part, the energy potential of the female floral meristem should be higher than the male one.

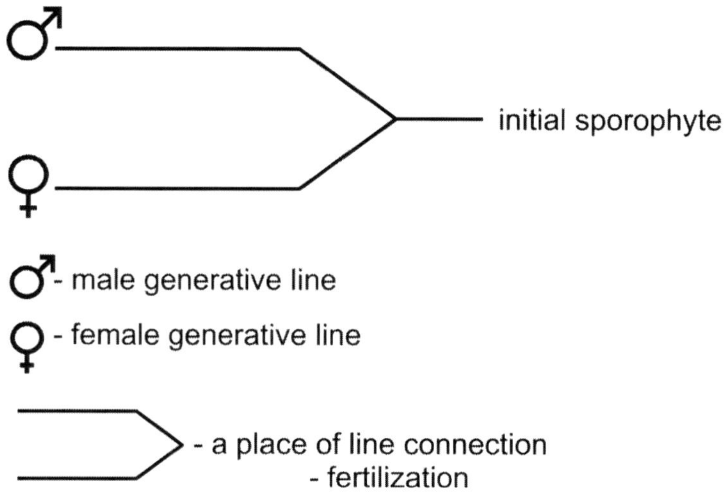

Figure 45. General scheme of a reproductive cycle.

During sexual development a pine as a monoecious plant gives rise to only the diploid bisexual sporophyte initials of two types (♀ ↑ + ♂ ↓ or ♀ ↓ + ♂. The differentiation of gametes by the sex occurs on the two stages of the reproductive cycle – microsporogenesis and megasporogenesis. The stage of microsporogenesis is located on a male generative line (♂), and the stage of megasporogenesis is on a female (♀) generative line. During meiosis at the microsporogenesis four morphologically identical male gametes - 2 gametes with a spin-up ↑ and 2 gametes with a spin-down ↓ - are formed.

The sexual system has passed to the next step of evolutionary development after the appearance of activating meiosis, which has entered into the reproductive cycle of all species of higher order plants. As it was already stated in chapter 3, the scheme activating meiosis includes the two different types of cell divisions – usual meiosis and activating mitosis, and directs the change of the DNA-particle and the DNA-wave of 4 gametes. The gamete, a DNA-wave of which is raised to higher energy levels, is viable due to the other three gametes, which lose biological energy, and therefore will soon die. Moreover on the basis of a female generative line of a reproductive cycle is created "a cell apparatus of Stern – Gerlach", which represents the three consecutive cell divisions – mitotic and two divisions of activating meiosis (see Chapter V).

In *Pinis sylvesteris* the spin direction of female generative cells is dependent on the spin direction of the spermatogenous cell and the filtration of chromosomes in "a cell apparatus of Stern – Gerlach". Its essence: if the spin of the spermatogenous cell is located ↑, then the spin of the ovicell will be directed ↓, and likewise to the contrary. In detail the mechanism of the definition of sex will be described in this chapter.

Stages of the Emergence of a New Reproductive Initial

It is noticed that the emergence of any reproductive initial is preceded by a formation of the specific morphological structures of the same type – the local structures having on a cut the form of an oval or a circle (Allen, 1946; Erspamer, 1952; Singh, 1978; Owens, Molder, 1979; Kozubov, etc., 1982; Scupchenko, 1985; Kuznetsova, 2002, etc.). The local structure begins being formed as a result of sharp increase of the mitotic activity of one (few) meristematic or generative cells.

Characteristic signs of this stage are - the periclinal divisions of nuclei during local structure formation occur in different directions. Therefore the structure quickly grows and keeps an oval or circle form. And the nuclear divisions can be accompanied by cytokinesis (generative primordium, archegonium) or the nuclei divide, and a cell envelope between them is formed later (female gametophyte, free-nuclear proembryo). When a growth of local structures comes to an end, a quantity of the cells in them is defined by a specific number (Singh, 1978; Kozubov, etc., 1982; Trenin, 1988). All cells of the structure are non-differentiated with the same ultrastructure.

Later on the development of local structures can go in two directions (Figure 46). First, among the cells of a structure one or several generative cells are chosen, which become the reproductive initials.

Only the initial participates in a reproductive process. Other cells of the structure, having executed a specific function (most often trophic function), die off. In this way the development of two generative primordium, female gametophyte, free nuclear proembryo occur (Figure 46-a, c, e). Second, each cell of a structure becomes the reproductive initial and the uniform reproductive process divides into the independent lines, which are of equal number to the quantity of reproductive initials. In this way the development of the sporogenous cells of microsporangium is carried out (Figure 46-d).

Let us consider the construction of the local structures, when their development occurs in the first way. The cells of the superficial layer of the structure are located radially. And if you were to draw a longitudinal axis of symmetry through each cell of this layer, then all the lines will meet at one point and will divide a semicircle of any local complex on the approximately identical sectors in the form of "fan" (Figure 47-a, c, e). I.e., the orientations of the longitudinal axis of symmetry in all cells are different.

As has been shown in the analysis, that those cells of a complex structure become the reproductive initial and the symmetry axis of which coincides with the gravitational direction. We suppose that the definite spatial position of cells in a superficial layer of a structure may transform them in the reproductive (male, female) initial, promoting the emergence of intracellular changes, which change further the destiny of these cells.

All reproductive initials of *Pinus sylvestis* have arisen in this way: microsporangium initial, ovule initial, and an archegonial initial and embryonic initial (Figure 46-b, d, f). On the scheme of a reproductive cycle we will mark the stages of emergence of a reproductive initial by means of the symbol → .

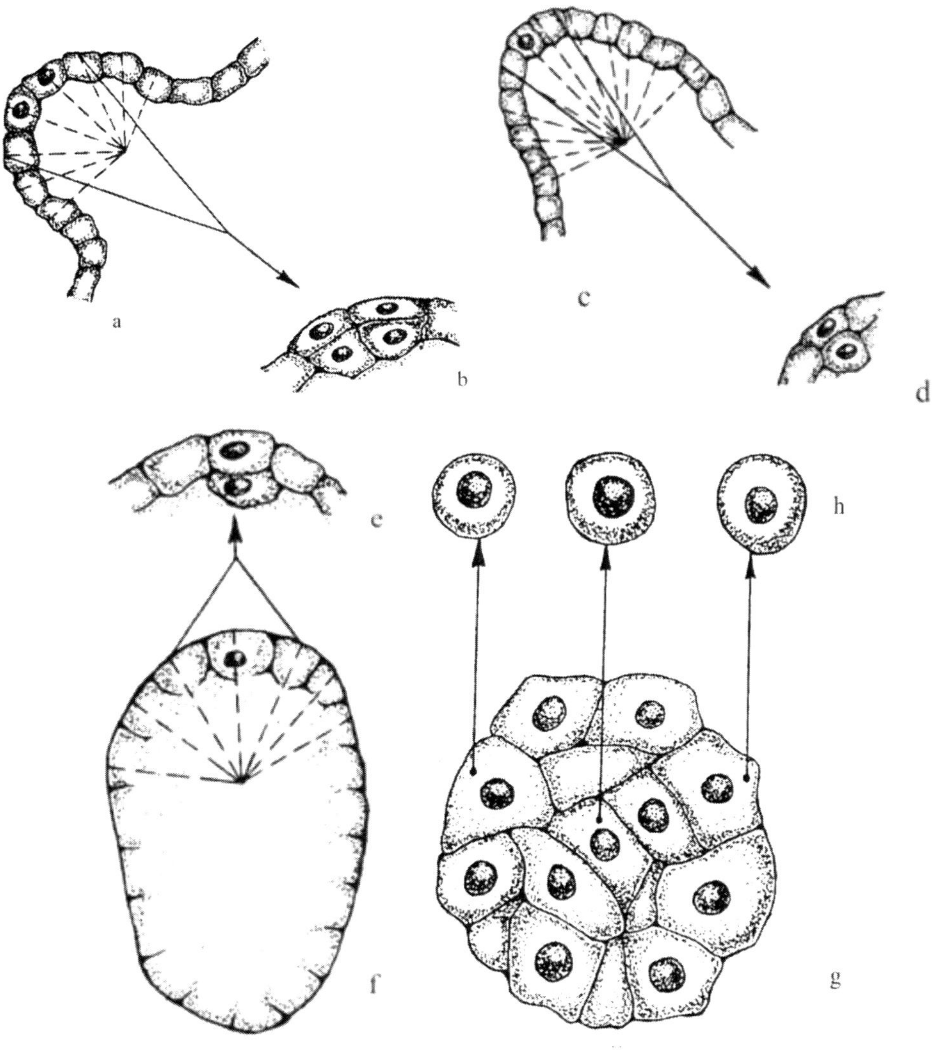

Figure 46. The stages of emergence of new reproductive initial in *Pinus sylvestris* L.a – primordium of microsporophyll; b – microsporangium initial; c – primordium of megasporophyll; d – ovule initial; e – female gametophyte; f – archegonial initial; g, h – sporogenus tissue of archesporium (g) and microspores (h).

Activation Stages or the Stages of the Creation of the Energy Program

The reproductive cycle of pine includes six activation stages. Each stage represents a part of the reproductive process, which is strictly organized in space and in time (Figure 47). With the first cell that divides, the activation stage begins, and it becomes a reproductive initial (microspore, ovule initial, archegonial initial, etc.). The stage comes to an end with the emergence of the apical cell. The transverse axial divisions between an initial and the apical cell are the main divisions and are presented in two types of divisions – activating mitosis and activating meiosis. The number of the main divisions varies from 2 (oogenesis, proembryogenesis) to 4 (microgametophytogenesis). The divisions of generative cells, which during activating mitosis (meiosis) move to a basal pole, are secondary. Each stage of activation is characterized by: 1) a strict sequence of cell divisions; 2) a constant number of the main and secondary divisions; 3) a fixed spatial condition of daughter cells in the planes of their divisions; 4) an identical number of cells, which form the initial and apical; 5) daughter cells that are the opposite by their predestination, destiny, metabolic and morphogenetic activity.

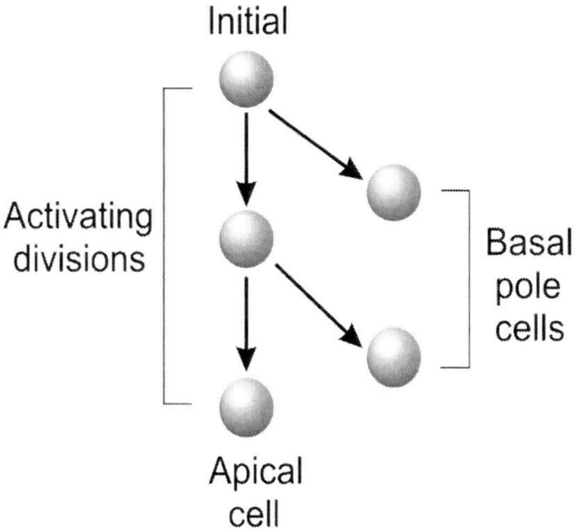

Figure 47. Scheme of structure of activation stage, which consists from a two activating mitosis.

Therefore, the main event, occurring during each activation stage, is the polar differentiation of daughter cells on viable active (or energetically strong) and apoptotic (or energetically weak) generative cells. All cells of the basal pole (prothallial cells, nonfunctional megaspores, etc. see table 2), as the energetically weak cells, leave a reproductive process. Their ability for division is limited. The cells of basal pole are not divided, or the number of secondary divisions is limited (usually 1-2 divisions). The apical cell, as the energetically strong cell, passes into the following reproductive stage (microsporocyte, megasporocyte), or participates in fertilization (male and female gametes), or forms an embryo (sporophyte initial).

The generative cells in the process of moving from one activation stage to another are changed essentially. They become larger, the polarity between the apical and basal pole has appeared, the nucleocytoplasmic ratio is altered (in a sperm and ovicell particularly), a high degree of chromatin dispersity is observed, the invaginations of a nucleus envelope in the increased nuclei of generative cells have occurred (Singh, 1978; Konar and Moitra, 1980; Kuznetsova, 1990, 1998, etc.).

According to the corpuscular-wave concept during the reproductive development of a sporophyte initial occurs as a supplement of the biological harmonious oscillator by the energy by means of the transition of a DNA-wave to higher energy levels in the six activation stages. The mechanism of the energy potential restitution is the transfer of biological energy from a chromosome to a chromosome during special cell divisions – an activating mitosis and activating meiosis. The process of "rejuvenation" of generative cells proceeds in definite physical conditions, which are created in the reproductive structures (microsporangium, ovule, female gametophyte, etc.). Under the law of energy conservation the energy potential of an apical cell increases as much as it decreases in the cells of the basal pole. The "rejuvenated" apical cell passes into the next reproductive activation stage where it receives one more portion of biological energy, and then on a following stage etc. Only the last generative cell (or sporophyte initial), which has the whole energy and hereditary program of development, can begin a new ontogenesis cycle. All generative cells of a basal pole, which during an activating mitosis loses biological energy, die.

Once again we can observe that all process of the formation of the energy program of a sporophyte initial is under strict genetic control. The experimental data testifies, that the number and sequence of the reproductive stages, which increase the energy state of DNA-wave, and the internal structure of each activation stage are the species-specific and conservative signs for all species. We will mark the activation stages on the scheme of a matrix of a reproductive cycle by means of the symbol → .

Table 2. The scheme of cell divisions during microgametophytogenesi, megasporogenesis, oogenesis and proembryogenesis as an example of reproductive cycle *Pinus sylvestris* L

Reproductive stage name; reproductive initial	N divi-sion	Cells of apical pole	Cells basal pole	Number divisions of basal cells	Scheme divisions
Gametogenesis Microspore	1	Antheridial initial	1st prothallial cell	0	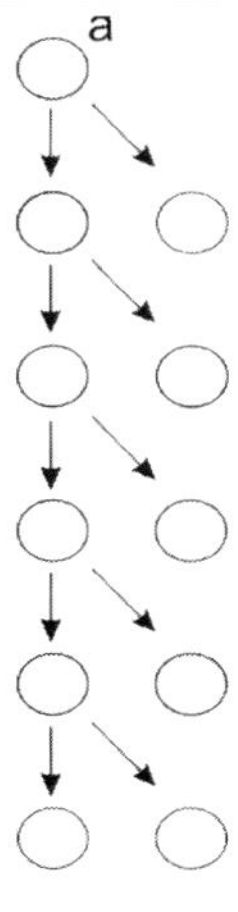
	2	Antheridial initial	2nd prothallial cell	0	
	3	Generative cell	Siphonogenous cell	0	
	4	Spermatogenous cell	Sterile cell	0	
	5	Functional sperm – participates in fertilization	Nonfunctional sperm	0	
Megasporogenesis Megasporocyte	1	Micropilar cell	Chalazal cell	1	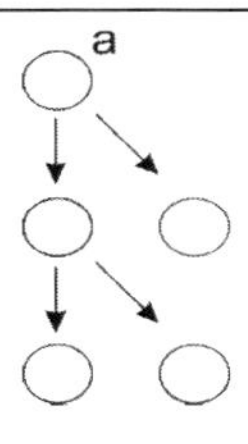
	2	Functional megaspore – gives rise to female gametophyte	Nonfunctional megaspore	0	

Reproductive stage name; reproductive initial	N division	Cells of apical pole	Cells basal pole	Number divisions of basal cells	Scheme divisions
Oogenesis Archegonial initial	1	Central cell	primary neck cell	2	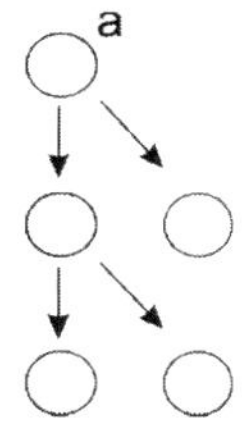
	2	Ovicell – participates in fertilization	Ventral canal cell	0	
Proembryogenesis 4—nuclear proembryo	1	primary embryo tier	Primary proximal tier	1	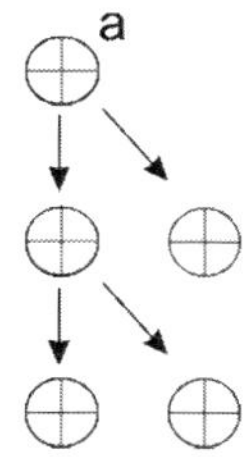
	2	embryo tier of 16-cellular proembryo – gives rise to an embryo	Suspensor tier	0	

[a] – reproductive initial, or the first cell, from which begins the activation stage.

Corpuscular-Wave Model of a Reproductive Cycle

In *Pinus sylvestris* the scheme of the matrix of a reproductive cycle has a typical forked form where one tooth of the "fork" is a male generative line, the second tooth is a female generative line, the place of their union is a fertilization, and the end of the handle is a sporophyte initial, which is at the same time the last generative and first vegetative group of cells (Figure 45). As shown in the analysis, a two complex biochemical and biophysical processes caused by the change of a DNA-particle and a DNA-wave and a formation of the new hereditary and energy program of a sporophyte initial on the basis of the matrix of a reproductive cycle proceed. As a confirmation of this we will consider the reproductive cycle of a pine from the above-mentioned positions.

Male Generative Line

According to experimental data the formation of a microstrobile begins with a formation of primordium of a microsporophyll (Allen, 1946; Owens, Molder, 1979, 1997; Kozubov et al., 1982; Bhatnagar and Moitra, 1996; Williams, 2009; etc.). The group of cells near the generative apex in a peripheral zone under the protoderm is isolated. They actively divide and give rise to a primordium. On a longitudinal cut it consists of the meristematic cells of the same type. The superficial layer of primordium represents the closed radial row of cells (Figure 46). A microsporangium initial become the group of cells from this layer (Fagerlind, 1961; Singh, 1978; Kozubov et al., 1982), at which the polar axis of symmetry coincides the closest with the direction of gravity. The described process exactly corresponds to the stage of the emergence of a new reproductive initial –. Therefore, the reproductive development of generative cells by the male generative line from this stage begins.

The reasons and mechanisms of the redetermination of an individual development path and the transformation of a meristematic cell in the generative cells are not known. Supposing, it is true that they are connected by the wave nature of a cell. In this case the biological harmonious oscillator in a life cycle should have two zones of effective frequencies, on which the transition of a DNA-wave to a higher energy level is possible. Due to that theory, the threshold level of biological energy for the male generative line should be lower, than that of the female generative line. Two facts testify to it. First, a development of a female generative sphere of pine begins 1-3 years earlier, than a male. Second, the megastrobils in a tree crown are located above, and the microstrobils are located below.

It is known that the course of the development of a microsporangium and its walls differ in their stability (Esau, 1980). The cytoembryology of microsporangiumgenesis in conifers is in detail described by Fagerlind (1961). Proceeding of an experimental data, the process of microsporangiumgenesis represents the first activation stage () by the male generative line and has the next structure. The cells of a microsporangium initial, which formed on the previous reproductive stage (), undergo an activating mitosis three times. During each transverse axial division the apical cells (one or few) inside a primordium are pushed through more deeply. The apical cell of this stage is the primary sporogenous cell, which passes on to a further reproductive process.

A primary sporogenous cell, unlike the meristematic cells of a primordium is already a generative cell, although a basis of heredity (a DNA-particle) in them is identical. It is characterized by a higher biological energy, which received this generative cell (on the basis of a DNA-wave) during microsporangiumgenesis. Then on the one hand one or few primary "rejuvenated" sporogenous cells, after a specific number of periclinal divisions, have formed. The process of microsporangium genesis is the first activation stage () on the male generative line. A primary sporogenous cells, as the energetically strong ones, begin to divide and form the archesporium. On the other hand three cells of the basal pole, which have arisen during the activation stage and have a less biological energy, undergo in this time the secondary divisions and give rise to the microsporangium.

Between the primary cells of archesporium and microsporangium, as between energetically weak and energetically strong reproductive structures, there is an energy gradient (see Chapter 6). That is why the derivatives of the three energetically weak cells of a basal pole are attracted to a sporogenous tissue, surrounded by radial ranks, forming the three layers of a microsporangium wall. The wall of a microsporangium ends its formation when a sporogenous cells transform into the microsporocytes and exists due to the epidermis, subepidermis, average layer and tapetum (Kozubov et al., 1982; Trenin, 1988). The tapetal cells in the contact area get to a zone of capture of in the summary biofield of an energetically strong cell of archesporium. The tapetum under it action is reorganized in a tissue of a secretory type and provides the generative cells of an expanding archesporium the nutrition. The cells of the subepidermis and average layers have a similar ultrastructure, but differ by the form and the size. During meiosis all layers, which have been generated by basal pole cells, are destroyed (Dickinson and Bell, 1976). The microsporangium after the pollen maturation has only the epidermal layer, which was formed from the superficial cells of a primordium of microsporophyll.

The sporogenous tissue is located in the central part of a microsporangium. In this period its cells are already easily identified (Figure 14). An archesporium represents the local group of the cells of the same type, which has the form on a cut of an oval. The sporogenous cells in comparison with the vegetative cells are larger in size the nucleocytoplasmic ratio in them has changed (Kozubov et al., 1982; Pennel and Bell, 1985; Trenin, 1988; etc.). They are characterized by the less electron-dense cytoplasm, a polygonal form that has thin and straightforward cell walls. A cytoplasm is poorly vacuolated and contains a great number of small plastids and mitochondrion.

The period of the archesporium formation is the second stage due to the emergence of a new reproductive initial () on the male generative line. This stage begins from the primary sporogenous cell (one or few) and ends by the origination of a great number of new reproductive initials – the maternal cells of microspores. Each reproductive initial independently enters into the next stage and the united reproductive process is separated on the independent (by the number of microsporocytes) developmental pathways.

As it has been stated in the previous chapter, the restriction stage of development is characterized by the following signs. During development of reproductive structures into the limited volume, as the result of an intracellular pressure increase, at the beginning the proliferous processes have ceased, then the growth of the generative cells stops and the generative cells pass to a differentiation and then to a rest state. The similar picture is observed during the transformation of sporogenous cells in the microsporocyte's (Owens and Molder, 1971; Singh, 1978; Trenin, 1986, 1988; Tretayakova, 1990; Williams, 2009; etc.). The development of archesporium occurs in the central part of a primordium of microsporophyll, which together with a formed wall of a microsporangium more and more prevents the divisions and the growth of sporogenous cells and gradually transits them to a restriction stage of development. According to the experimental data during maturation a microsporocyte's become larger, than the cells of the archesporium. Then their growth ceases and the accumulation of starch grains and lipid drops begins. It is necessary to notice, that the higher pressure is the main physical condition for the initiation of meiosis as well. Therefore during microsporogenesis a microsporocyte undergoes the meiotic division.

On a male generative line the usual (or male) type of meiosis takes place. During meiosis the 4 haploid microspores are formed. These gametes are the totipotent cells, because they have an identical DNA-wave and during gametogenesis the same cell lines formed from them. On the other hand, these gametes differ on a basis of heredity (a DNA-particle) and by a spin direction (two gametes is ♂↑ and two gametes is ♂↓). A DNA-particle and the sex of gametes change by means of different quantum-mechanical mechanisms – the crossing-over and filtrations of nucleotides in a non-uniform magnetic field. Both these mechanisms are a part of the scheme of meiotic division and proceed independently from each other. Thus, the microsporogenesis is a stage of formation of the new hereditary program () and the definitions of the sex of gametes (⚥). A combination of several processes within one stage on the scheme of the reproductive cycle is marked by this symbol"+".

After a change of a DNA-particle during crossing-over between microspores and a tapetal tissue, as between two genetically heterogeneous and energetically different systems, the processes of incompatibility arise. Therefore in diplotene a callose envelope around the microspores begins to form, which progressively thicken (Trenin, 1988). This layer of callose is the first abscission layer which is created during microsporogenesis. We think that such a powerful callose "case" and the spatial separation of microspores into tetrads should loosen the action of attraction forces between two gametes in the combination ♂↑ + ♂↓ as well. Free energy, which during decondensation of a meiotic chromosome enters twice into the volume of the generative cells, causes the intensive growth of microspores and the destruction of their callose envelopes. However before its destruction the new envelope of a pollen grain between a callose layer and a plasmalemma begins to form (Singh, 1978; Kozubov et al., 1982; Trenin, 1986, 1988; Tretayakova, 1990; Williams, 2009; etc.). A sporoderm is the next abscission layer, which limits the growth of male gametes, initiates a transition of microspores in the next activation stage and protects a maternal plant from the high biological energy of the pollen grains.

A sporoderm represents the external envelope of a pollen grain, which exists due to the two layers – an intine and an exine (Figure 41-a). The main chemical component of this envelope is the sporopollenin, one of most firm biopolymers in the organic world. A sporoderm, as the effective and stable abscission layer, during its formation begins to limit the growth of microspores and isolates them from each other. The growth in a fixed space causes an increase of intracellular pressure into each generative cell. The raised pressure initiates the transition of gametes to the activating mitosis and simultaneously to the restriction stage of the development. Therefore after the first activating mitosis large starch grains and numerous lipid drops have appeared (Trenin, 1986, 1988). After the second division a cytoplasm is filled with a protein bodies. After the third division the process of progressive dehydration begins and the transition of pollen grains into a rest state.

The gametogenesis (or microgametophytogenesis) is the second activation stage (⊧) by a male generative line (see the table). During gametogenesis a microspore undergoes 5 divisions, four of which are realized inside the spores and present the activating mitosis. As a result of 5 activating mitoses on the basal pole a number of unviable cells are consistently formed: two prothallial cells, siphonogenous or tube cells, sterile cells and nonfunctional sperm. On apical pole of this stage the functional sperm is formed, which participates in fertilization. After gametogenesis the biological harmonic oscillator of a given sperm is the most energetically strong structure in the male generative line. It is necessary to notice, that only a siphonogenous cell forms a pollen tube out of four cells of the basal pole. The other 3 cells (2 prothallial cells and 1 sterile cell), having given all biological energy and having gone down to the zero energy level, soon die. Therefore we have a basis for the assumption that the male generative line, having begun development with a lower energy level, in comparison with a female generative line, has risen to a higher level before the pollen germination.

The mature pollen grain contains 4-5 cells: the remains of two prothallial cells, a large siphonogenous cell and a small generative cell, which undergoes the fourth division before or during pollination (Nekrasova, 1983; Trenin, 1986; Owens and Bruns, 2000). The last division occurs in the next year before fertilization and leads to the formation of two male gametes (sperms). During reproductive development the degree of chromatin condensation in the nuclei of male gametes has changed. A chromatin of sperm is presented in the euchromatin, which is dispersed in the nucleus volume, and heterochromatin is absent.

Thus, development of the structures of a reproduction in a male generative line includes 6 stages: one stage by the creation of the hereditary program (⨳); one stage by the definition of gamete sex (⚥), which connects with the previous stage; two activation stages by the creation of the energy program (ᚦ), and two stages by the emergence of a new reproductive initial (⬭), which connects with the activation stages. The scheme of the disposition of stages on this line is shown in figure 48.

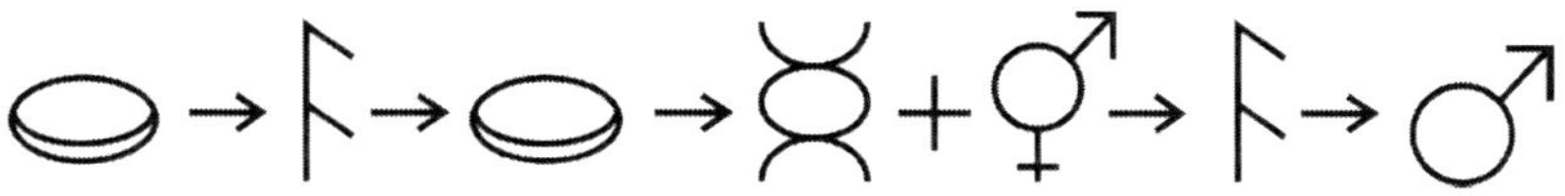

Figure 48. Scheme of stage succession of male generative line of the *Pinus sylvestris* L.

Female Generative Line

In conifers a development of primordiums of megasporophyll of the same type occurs (Singh, 1978; Kozubov et al., 1982; Trenin, 1986, 1988; Tretayakova, 1990; Van Vent and Williams, 1990; Bhatnagar and Moitra, 1996; Williams, 2009; etc.). It is the non-differentiated meristematic cells. The cells of the external layer are densely pressed to each other and represent a radial row (Figure 46-c). The ovule initial become a superficial cell of the primordium (Brunkener, 1973). The described process corresponds to the stage of the emergence of a new reproductive initial (⬭). A formation of the new hereditary and energy programs of future plants by a female generative line begins from the given stage.

During megasporangiogenesis an ovule initial is twice divided by means of activating mitosis. The external derivatives of these divisions, having passed the fixed number of secondary divisions, form the two layers of the megasporangium wall – integument and nucellus. The apical cell of transverse divisions is the primary sporogenous cell, which becomes a maternal cell of the megaspore (Singh, 1978; Kozubov et al., 1982). During activating mitosis a primary sporogenous cell is carried over into the center of the primordium, where there will be a further development of generative cells. It seems obvious, that the process of megasporangium formation is the activation stage (ᚦ) – the first on a female generative line. A primary sporogenous cell, having received a portion of biological energy during megasporangiogenesis, becomes the first generative cell. During megasporangiogenesis a redetermination of the developmental pathway (on pathway of sexual reproduction) occurs, as the result of a DNA-wave transition of an ovule initial (a vegetative cell) to higher energy levels.

After that a primary sporogenous cell begins to increase in size, undergoes 2-3 mitotic divisions and gives rise to 4-8 sporogenous cells (Figure 49-a). The process of the formation of the sporogenous cells is the second stage that occurs due to the emergence of a new reproductive initial (⬭).

The ovule, before meiosis has the multicellular nucellus and integument and the only increase in size of a maternal cell megaspore (Figure 49-b). However the details of this process, as well as the microsporangiogenesis, are not studied: there are no names for the first cells of a basal pole, the number of the secondary divisions is not known, etc. Therefore these stages are not evidenced in table 2. We suppose that the tissues of nucellus and integument have arisen from the primary cells of the basal pole, which during megasporangiogenesis have lost the majority of their own biological energy. According to the 2nd thermodynamic law these cells and their derivatives will continue to go down from level to level up to the zero energy level and form the two types of energetically weak cells. At this time a primary sporogenous cell (i.e., the apical cell of megasporangiogenesis, which receives this portion of biological energy) and its derivatives, as energetically strong cells, begin to attract the nearby energetically weak cells and form a layer of nucellar tapetum (Figure 49-a, b).

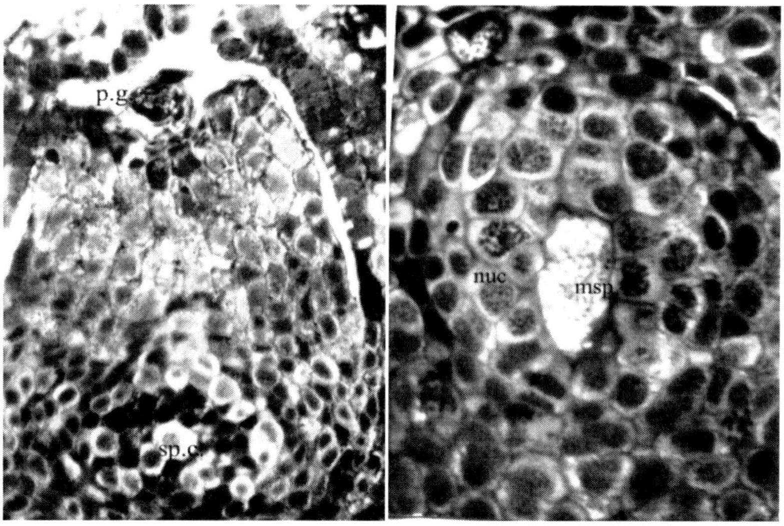

a b

Figure 49. Apical part of nucellus of ovule and sporogenous cells of *Pinus sylvestris* during pollination (a) and maternal cell megaspore before the first meiotic division (b).p.g. – pollen grain; sp.c. – sporogenous cells; nuc – nucellus; msp – megasporocyte.

During the analysis of reproductive cycle of the pine we have paid attention to several facts, which are explained on occasion, if a generative cell represents the oscillatory system. 1) During pollination into the pollen chamber, as a rule, gets from 1 to 4-5 pollen grains (Figure 50-a). After germination they give rise to the pollen tubes, which in the 1st year of development compete among themselves (Figure 50-b). However in the 2nd year of development only one pollen tube renews its growth and only one nucleus of the sperms participates in fertilization (Figure 20-b). 2) A vestige of the leading pollen tube which is left in a nucellus after fertilization represents a wide straight line. 3) If after pollination the pollen in the pollen chamber is absent the meiosis stops during first division. Without pollination the ovules are degraded during 2-3 weeks (Sarvas, 1962; Chromova, 1985; Tretayakova, 1990). 4) The early stages of the origin and development of the primary sporogenous cells and ovules in gymnosperms are not studied in detail in spite of the numerous investigations.

The anatomic structure of the ovule and reproductive processes is well traced only since the megasporogenesis stage. 5) It is especially interesting to notice, that the first meiotic division occurs only after pollination or soon after it (Kozubov, 1973; Trenin, 1988) (Figure 50-d). It is considered, that a pollen tube growth carries out chemotropically by means of the substances of a synthesized ovule (Herero, 2001). All this data leads to the conclusion, that the origin of a maternal cell megaspore from one of the sporogenous cells is somehow connected with the appearance into the pollen chamber of pollen grains.

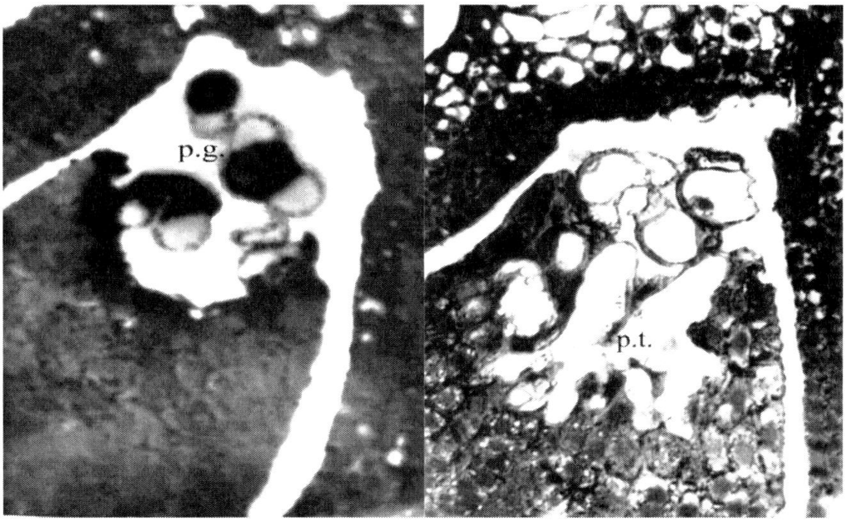

Figure 50. 4 pollen grains *Pinus sylvestris* into pollen chamber (a); 3 pollen tubes in a month after pollination (b) and pollen tube during fertilization (c); prophase meiosis (d).p.g. – pollen grain; p.t. – pollen tube.

During germination a pollen grain has only two living cells – an energetically strong generative cell and an energetically weak siphonogenous cell. The biological harmonious oscillators of both of these generative cells possess a high quality (Q) and for a relatively long time the constant amplitude of DNA-wave oscillations is supported on every energy level. However, as the siphonogenous cell possesses a low energy potential, we will not take it into consideration. A generative cell after the activation of 2 stages is the most energetically strong cell of male generative sphere. We think that its energy potential above that in female sporogenous cells which up to the beginning of pollination have passed only one stage of activation. Therefore if we proceed from the quantum-mechanical nature of male and female generative cells and experimental data it is possible to assume the following hypothesis of the biophysical nature of megasporogenesis and the role of the wave properties of pollen and the female sporogenous cells in their development on the given part of a reproductive cycle.

Supposing a few pollen grains get into the pollen chamber. Each generative cell of a pollen grain represents the independent oscillatory system, which possesses a specific resonant frequency $\omega_{♂res.}$ Accordingly the female sporogenous cells are also 4-8 independent oscillatory systems with a spectrum of resonant frequencies $\omega_{♀res.}$ Therefore the biological harmonious oscillator of each pollen grain interacts simultaneously with all 4-8 harmonic oscillators of the sporogenous cells. However according to experimental data the end result of the interaction of these generative cells is a transition in a reproductive process of only a sporogenous cell – the future megasporocyte. All other sporogenous cells die soon after the transition. Perhaps, it is possible to explain the physical nature of this phenomenon as follows.

Now it seems obvious, that the biological harmonious oscillator of the pollen, which first reached the nucellus, interacts with the oscillators of 4-8 female sporogenous cells, whose frequencies somewhat differ from each other. First of all we will note, as the biological harmonious oscillator of all cell systems it is characterized by high quality (Q), their resonant curve should be with a very sharp maximum. The phenomenon of a resonance or the increase of oscillation amplitude up to the maximum occurs, if the resonant frequencies of the DNA-wave of a pollen generative cell ($\omega_{♂res}$) and female sporogenous cell ($\omega_{♀res}$) coincide. Accordingly, the response of sporogenous cell due to the resonance will be considerable. It directs the essential increase of its energy potential and up to the essential change of the internal organization of the ovule as a whole. The activated sporogenous cell begins to quickly grow and soon changes into a maternal cell megaspore. Therefore the megasporocytes are easily identified on the micrographs: they are larger than the cells of the primordium, the nucleocytoplasmic ratio is changed, the cytoplasm polarity takes place, etc. (Konar and Moitra; 1980; Singh, 1978; Kozubov et al., 1982; Trenin, 1986, 1988) (Rice 49-b). During the resonant response of a non-activated sporogenous cells, energetically weak cells of nucellus and integument are attracted towards the megasporocytes. As a consequence, the ovule gets a typical anatomic structure for the given stage.

Really, the formation of the megasporocyte is preceded by 2-3 mitotic divisions of sporogenous cells (Singh, 1978). A resonant response and interaction of male and female generative cells are confirmed by the intensive growth of one sporogenous cell and the appearance of a maternal cell megaspore. A DNA-wave of this activated generative cell is characterized by the maximum amplitude of oscillation and, accordingly, a very high energy. Apparently, the biological harmonic oscillator of a generative cell of pollen supports the maximum level of the amplitude of the DNA-wave oscillation of a megasporocyte during the entire time of its maturation. Therefore the biofield of a megasporocyte is capable of changing the spatial orientation of the other sporogenous cells, a nucellus and integument tissues, and their further development of subordinates to itself. The polarity of a megasporocyte and the change in a direction of a karyokinetic spindle during activating meiosis may be the result of the powerful resonant response of a sporogenous cell as well.

It is necessary to notice that the non-activated by first pollen grain sporogenous cells, as the oscillatory systems, cannot interact with other pollen grains in a pollen chamber, because they entered into the zone of the very powerful biofield of the activated megasporocyte. Therefore only the leading pollen tube, which has the connection with the biological harmonious oscillator of a megasporocyte, participates in the fertilization, in spite of the fact that the main part of pollen grains germinates on a nucellus and forms the normal pollen tubes. Apparently, the mechanisms of pollen germination on the nucellus and on the nutrient medium are the same. As to the incompatibility processes (for example, during inbreeding), they can stop the growth of the leading inbred pollen tube and fertilization in this case does not occur. As known, self-pollination leads to the partial reduction of the cone number and the development of empty seeds in *Pinus sylvestris* (Kuznetsova, 1996).

Before fertilization a megasporocyte presents the large formative cell of oval form, located in the center of the ovule. Therefore its growth is suppressed by the tissues of the nucellus and integument, whose cells closely adjoin to a maternal cell megaspore and to each other. It is necessary to specify also, that a pine has the nucellus of a crassinucellar type. The cells of its internal zone, which turn out into the area of the powerful biofield of megasporocyte, are transformed in the secretory cells of a nucellar tapetum with a specific ultrastructure. The cells of a nucellar tapetum, supplying a megasporocyte by the nutrients, are destroyed gradually and replaced by a following layer, and then the following.

As it was already spoken, a growth and division of cell and cell systems into the limited volume leads to the increase of intracellular pressure and to their transfer to a restriction stage of development. Therefore the growth of the megasporocyte gradually stops, a cytoplasm accumulates the nutrients (starch and lipid drops) and the cell enters into a meiosis.

At the time of the analysis megasporogenesis we have noted, that the spatial orientation of generative cells and their division spindles during three successive divisions – a mitosis and two meiotic divisions resembles the well-known experiment of a filtration of atoms by means of the three apparatus of Stern-Gerlach for the Fermi particles with spin of ½ (see Chapter 5, Figure 35). Its essence: if the last two apparatus together unfold on 90^0, from this physical system only half of the particles of one kind terminate (Feynman et al., 1963). I.e., if a cell is the quantum-mechanical system, the cytoembryological data testify to the possibility of the existence of "a cell apparatus of Stern – Gerlach". We assume that during resonant interaction between the male and the female generative cells and on the basis of "a cell apparatus of Stern – Gerlach "the physical essence of a megasporocyte changes. A functional megaspore becomes a polarized cell with a specific direction of spin. We will look at the physical nature this process in more detail.

In figure 35 the disposition of the karyokinetic spindles during the three consecutive divisions of megasporogenesis is shown: at the time of the last mitosis of a sporogenous cell before a meiotic division; and the two divisions of an activating meiosis. If nucleotides are Fermi-particles, through the system of "a cell apparatus of Stern – Gerlach "will pass particles either with the spin ↑ or with the spin ↓. It has been asked, what kind of particles on the exit of system should be formed?

The peculiarity of the interaction of the two analyzed oscillatory systems is the following. The generative cell of pollens is the polarized cell (i.e., after meiosis has a specific direction of spin), and the female sporogenous cell is the non-polarized. Therefore we can assume that during the first reducing division of an activating meiosis, when a DNA-wave of a megasporocyte passes through the sharp resonant chain of the frequency of the male generative cell, the physical essence of micropylar and chalazal cells, as the non-polarized cells of the apical and basal poles, are defined. According to the Pauli exclusionary principle, the spins of micropylar and spermiogenous cells, as the two energetically strong generative cells, are arranged towards each other. The spin of the energetically weak chalazal cells, which has the opposite direction of spin from the micropylar cell, should be in the identical direction with the generative cell of pollen. I.e., if the spin of the male generative cell is situated ↑, then the spin in a micropylar cell is ↓, and, respectively, in a chalazal cell is ↑; and likewise to the contrary. During the secondary meiotic division an analogous situation takes place, because at the time of resonant interaction between a functional megaspore and a generative cell of pollen (energetically strong cells) their spins should accept the oppositely directional spins as well.

Thus, in conifers the sex of the functional megaspore is defined by the spin direction of the pollen grain, which first reaches a nucellus and begins to interact with the biological harmonic oscillator of a female sporogenous cell. It follows therefore that a megasporogenesis is the stage of the definition of the sex of functional megaspores (⚥) on the female generative line. From the aforementioned it become obvious, since the sex of a functional megaspore is determined by the spin direction of the pollen, then the sperms and ovicells with opposite spins only should participate in the process of fertilization.

A maternal cell megaspore undergoes the two divisions of activating meiosis. During crossing-over, which leads to the recombinational variability of a gene, a DNA-particle alters and a new hereditary program by the female generative line begins to form. Therefore, a megasporogenesis is also the stage of the change of the hereditary basis of the haploid megaspores ().

In chapters 3 and 5 we pointed out, that the process of activating meiosis (or the female meiosis) unlike the male meiosis in a space is organized in the other way (Figure 35). A polar axis of a megasporocyte and the karyokinetic spindles of two meiotic divisions coincide with each other (Juel, 1900) and with the direction of gravity. Therefore, if we look at the figures of the meiotic divisions and compare them with the scheme of an activation stage which consists of two activating mitosis (Figure 47), it is clear, that they are identical. After each division of meiosis towards the basal pole by one energetically weak impractical cell accordingly, a micropilar cell and a nonfunctional megaspore terminate (Table 2). An apical cell of a megasporogenesis is the energetically strong functional megaspore, which continues the reproductive process by a female generative line. Therefore the meiosis at megasporogenesis is the typical activation stage or the next stage due to the creation of the energy program of a future plant ().

In comparison with a maternal cell megaspore the ultrastructure of a functional megaspore has essentially changed. It is characterized by larger sizes. The volume of the nucleus on the one genome has sharply increased, because a nucleus becomes larger and the chromosome number is decreased by half. The quantity of heterochromatin in a nucleus has decreased. However the internal essence of the genome of a functional megaspore has been greatly changed. Megasporogenesis is the unique reproductive stage, whose role in the processes of sexual reproduction is studied only partially. According to our notions of this stage the three mechanisms of the change of a corpuscular-wave nature of a genome functional megaspore are realized: 1) a growth of the energy potential on the DNA-wave basis; 2) an alteration of the hereditary program on the DNA-particle basis and 3) a definition of the sex on the DNA-wave basis. On the scheme this combination of different processes in one reproductive stage is marked by the symbol "+" as well.

A functional megaspore is the first cell of a future female gametophyte. After meiosis this generative cell without a rest period begins to be divided. In the beginning its nucleus undergoes a definite number of free-nuclear mitosis, because the nuceocytoplasmic ratio of megaspore is strongly changed. As a consequence a free-nuclear female gametophyte is formed. At this time the female gametophyte, as the energetically strong structure, quickly grows and takes the form of an oval. A vacuole appears after the first megaspore division and gradually increases in size. The nuclei are built along an envelope and they are surrounded by a thin layer of cytoplasm.

The formation of the female gametophyte occurs in the center of the ovule. Therefore, its growth is restrained by means of a sporoderm and the so-called spongy tissue whose cells are disposed by the dense concentric rows. The beginning of a sporoderm formation is the free-nuclear stage and its termination is the cellular stage of a gametophytes development. At first the growth of a female gametophyte goes quickly, and then in the course of the sporoderm formation is gradually slowed down. It leads to a normalization of the nuceocytoplasmic ratio and to a transition from the free-nuclear to the mitotic divisions. Cell envelopes between the nuclei are formed centripetally; therefore the cells in a superficial layer represent a radial row (Figure 46-e). These first cells of endosperm have a strongly vacuolated cytoplasm with the identical ultrastructure (Figure 19-a). An archegonium initial becomes one (or a few) cells from the superficial layer of the micropylar zone of gametophyte, whose longitudinal axis of symmetry coincides with the direction of gravity. Therefore a megagametophytogenesis is the 3rd stage by the emergence of a new reproductive initial () in the female generative line.

The situation, which turns out between the tissues of a nucellus and a female gametophyte, can be considered as the interaction between two heterogeneous systems, because during meiosis a DNA-particle and a DNA-wave of a functional megaspore were altered. First, a DNA-wave passed to a higher level and the gradient between the energy potentials of these two tissues increased. Therefore the powerful biofield of a megaspore, and, accordingly, a female gametophyte causes a transition of the nearest cells of a nucellar tapetum into a very intensive regime of functioning, creates a centripetal stream of nutrients towards a growing gametophyte, attracts and then destroys the layers of the energetically weak nucellar cells up to the complete destruction of the nucellus tissue (see Chapter 5). Second, during crossing-over the DNA-particle chemical formula alters and that leads to change of the protein component of cell at all levels of their structural organization. Therefore a sporoderm origination is possible to explain by the action of a genetic system of incompatibility, which after meiosis begins to act on the boundary between a female gametophyte and a nucellus (Kuznetsova, 1996). The genetics and physiology of incompatibility processes is mostly studied in flowering plants (Heslop-Harrison, 1978; Pandey, 1980; Dickinson, 1990; etc.). The mechanisms of tissue incompatibility and spatial isolation of the gametophyte, apparently, are connected with the intensification of the processes of molecular selection. Naturally, not all biomolecules and ions, moving from the numerous destructing cells of a nucellar tapetum towards the gametophyte, can pass through a plasmatic membrane, which is characterized by selective permeability. Part of them (polysaccharides, some proteins, complicated lipids, etc.) are detained, accumulate and by means of the forces of a powerful biofield the identical molecules of the polysaccharides begin to polymerize, and, thus, a special stable abscission layer of sporoderm is forming. During thickening of the sporoderm the centripetal streams of nutrients gradually reduce, a growth of the female gametophyte slows down, then ceases, and its spatial isolation occurs. The development of a gametophyte inside the fixed volume leads to a rapid growth of intracellular pressure, to a termination of cell divisions, to an accumulation in the endosperm of nutrients (proteins, lipids) and to a transition of cells into a rest state (Figure 19-b).

During analysis of oogenesis we have paid attention to the interesting coincidence. According to experimental data 1-4 archegonium in a female gametophyte can develop (as a rule, 1-2; and is more rare at 3; and is absolutely rare at 4 archegonium forms) (Sarvas, 1962; Singh, 1978; Konar and Moitra, 1980; Kuznetsova, 1990; etc.). On the other hand the 1-4 pollen tubes in a nucellus also is formed. It has set us thinking, that the development of the archegonium is regulated by the generative cell of the pollen as the oscillatory system also.

It is established, that the most intensive growth of a female gametophyte takes place during the free-nuclear stage (Kozubov et al., 1982). The ovules spend the winter at the stage of 32 free nuclei (Fergusson, 1904). In the spring the division of nuclei again renews, and in beginning of the cellular stage their number becomes ≈ 2000. The volume of a female gametophyte in comparison with the volume of a megaspore increases more than 10000 times. Let us suppose now, that at the end of an intensive growth the apical part of a gametophyte reaches the zone where forces of a spin-spin attraction between a male and female generative cell begins to operate. From the side of a male generative sphere 1-4 generative cells with pollen tubes participate in a spin-spin interaction. From the side of a female generative sphere it is just a few superficial cells of endosperm from the apical zone of a female gametophyte. This interaction gives the powerful impulse to the development both the female gametes and the male gametes. The 1-4 endosperm cells become the archegonium initials and pass to the following reproductive stage. The 1-4 generative cells by means of the forces of spin-spin interaction are pushed out from the pollen grain and are delivered to the ovicell. The ovicell remains without fertilization, if there are not enough forces of attraction, and a generative cell remains in the pollen grain.

It is necessary to note, that under the probability theory the generative cells of pollen can have two spin directions – or a spin-up or a spin-down. An endosperm cells are polarized as the functional megaspore, which gives rise to a gametophyte. The one combination of ♂ and ♀ gametes with the opposite spins (↑↓) is certainly realized, because it has been created during megasporogenesis. The probability, that all male gametes will have the identical spin direction, is slight. Therefore 1-2 archegonia (3 rarely) participate in the fertilization.

Now we will move on to the analysis of oogenesis. During oogenesis an archegonium initial undergoes two transverse divisions of activating mitosis (Tablee 2, see). As a result of these divisions to the basal pole 2 nonviable energetically weak cells (a primary neck cell and a ventral canal cell) go away. On the apical pole is formed the ovicell, which participates in the fertilization. Therefore the oogenesis is a 3rd activation stage (ᚨ) and the last reproductive stage by a female generative line. During this stage an ovicell becomes the most energetically strong cell of the female gametophyte. A central cell, having received a portion of the biological energy, goes into a period of fast growth, vacuolation and acquires the look of a foamy structure (Figure 51-a). In this time a nucleus is disposed under the neck cells. A chromatin, as well as in the sperm nucleus, is presented by the form of euchromatin only. The neighboring cells of endosperm are attracted to a central cell just as to the energetically strong cell. Under the influence of the powerful biofield they transfer to a regime of intensive functioning and begin to carry out a trophic function.

The development of endosperm occurs inside a female gametophyte and represents the ordered reproductive process (Singh, 1978; Tretayakova, 1990; Williams, 2009). The endosperm unlike a nucellus is characterized by the higher energy potential. This tissue originates from a functional megaspore, which has passed through 2 activation stages (megasporangiogenesis and megasporogenesis). Therefore the gradient between the energy potentials of the endosperm and an ovicell arises only during oogenesis. The archegonium after reception of a portion of the biological energy begins to grow and during maturing its volume increases more than in 100 times. The development of the archegonium proceeds into a female gametophyte which has ceased growth, because it is surrounded outside by a very firm sporoderm. The cell of an endosperm has its growth restrained as well. Therefore the archegonium under the conditions of constantly increasing pressure passes to a restriction stage of development. On this path in the beginning the vacuoles disappear, the cytoplasm condenses; the central cell undergoes a second division of an activating mitosis and gives rise to an ovicell. Then the nucleus of the ovicell passes 4 divisions of endomitosis. As a consequence, the DNA quantity in an ovicell is increased 32 times (Ermakov, etc., 1981; Barantseva, 1992), the nucleocytoplasmic ratio, ultrastructure and chemical compound are changed as well. The growth of an ovicell is gradually slowed down and stopped. Then a cytoplasm begins to accumulate the nutrients, which are deposited in the system of large and small inclusions (Figure 51-b).

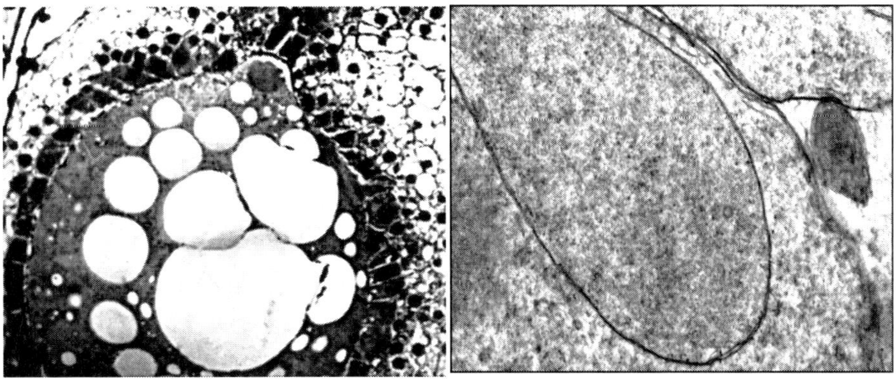

a b

Figure 51. The central cell (a) and fragment of ovicell cytoplasm before fertilization (b) in *Pinus sylvestris*.

We already stated earlier, that in the beginning of a cellular stage the forces of a spin-spin attraction between each archegonium initial and the generative cell of germinated pollen, apparently, start to operate. At the beginning the forces of mutual attraction are not enough in order for a generative cell to pass from a pollen grain into the pollen tube. After the 1st activating mitosis of oogenesis the archegonium initial gives rise to the central cell. The energy potential of the central cell is above and the biofield, accordingly, more powerful. However, in spite of the fact that the generative cells are mutually attracted more intensively to each other, the generative cell still stays in a pollen grain. The position of a nucleus which is displaced to an apical pole of a central cell can testify about the existence of spin-spin interaction. The ovicell after the 2nd activating mitosis of oogenesis is formed. It is the most energetically strong cell of the female gametophyte. In turn the generative cell is divided on the energetically strong spermatogenous cell and energetically weak sterile cell (Konar, Oberoy, 1969). Now that the male and female gametes are attracted very powerfully, the nucleus of the male gamete is capable of destroying the external wall of a pollen grain and the content of a spermatogenous cell go out in a pollen tube. In this time the male gamete quickly moves towards the ovicell. The 5th and last activating mitosis of gametogenesis occurs in the pollen tube before fertilization. The apical nucleus of this division is a sperm of greater size. The ovicell at this time undergoes 4 divisions of endomitosis and enters into a restrictive stage of development and a cytoplasm is filled with nutrients. A nucleus after the 2nd division of oogenesis in the center of the ovicell is situated and remains in this position up to the fertilization.

The pollen tube comes into contact with a female gametophyte, in which already are formed an archegonium with an ovicell (Trenin, 1988). The forces of a spin-spin attraction keep moving a functional sperm in the direction of a nucleus of an ovicell. Besides during maturation an ovicell, as the oscillatory system, begins to pass in the equilibrium state at a critical point, or to the fertilizable state. The ovicell in this state possesses a raised receptivity. Apparently, it is caused by the specificity of the physics of the equilibrium state, at which the biofield is concentrated mainly in a nucleus. Therefore the ovicell biofield cannot prevent a rapprochement of the gametes and a penetration of the energetically strong sperm into a cytoplasm, and then into a nucleus. Naturally, all ovicells without fertilization are doomed to death, because they are not capable from a balance go out independently. The destruction of a non-fertilized archegonium occurs in 7-10 days after fertilization.

Thus, the female generative line consists of the 8 reproductive stages: one stage by the creation of the hereditary program (), 3 activation stages, 3 stages by the emergence of a new reproductive initial () and 1 stage by the definition of sex (). The main biophysical characteristic feature of the given line: the processes of megasporogenesis and oogenesis are initiated and realized due to pollen participation, as an oscillatory system. The scheme of the disposition of stages on a female generative line is as follows (Figure 52).

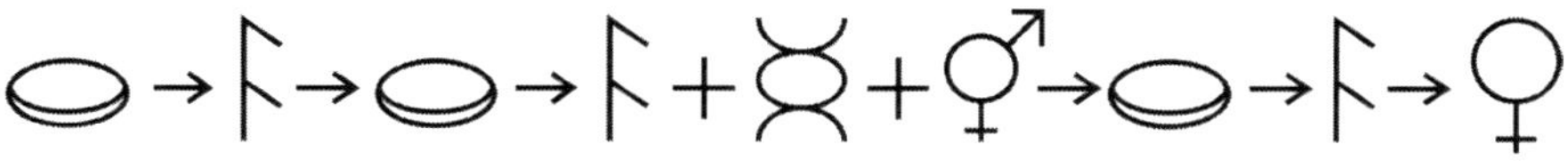

Figure 52. Scheme of stage succession of female generative line of the *Pinus sylvestris* L.

Fertilization and Proembryogenesis

The two most energetically strong gametes by a male and female generative line participate in fertilization give rise to a zygote. According to experimental data the fusion of a nucleus of a functional sperm with a nucleus of an ovicell occurs in the center of the female gamete and lasts for approximately 14 hours (Ermakov et.al., 1981). The amalgamation of the chromosome sets before the first division of a zygote does not occur (Singh, 1978). In metaphase, the chromosomes are split longitudinally and the two halves of the sets move to opposite poles. However, the truly diploid genome is absent both in a zygote, and in the nuclei of a 2-nucleus proembryo. The impossibility of the appearance of a truly diploid condition in this period is caused by the following reason.

During penetration of a sperm into an ovicell the genetic material in a zygote nucleus is united, which differs in corpuscular-wave characteristics. The ovicell is a very large immovable cell, which is in a resting state. Therefore a DNA-wave radiation and a biofield are absent. In a mature ovicell the volume of the nucleus which falls on the one genome is 2.5 times more than in a young central cell (Ermakov et al., 1981). A DNA-particle represents cumbersome polytene chromosomes with the DNA quantity of 32C (Ermakov, 1980, 1981, 1982; Barantseva, 1992). On the contrary, a haploid nucleus of a sperm has the small size with the DNA quantity of C. Their own cytoplasm is absent. On the physical essence, a sperm is the "charged" high-frequency quantum harmonious oscillator, which moves towards an ovicell under the influence of the forces of a spin-spin attraction.

The sperm penetration into a cytoplasm and nucleus of an ovicell changes the environment and the ovicell comes out of a rest state. The genome of the female gamete is destabilized more after a DNA-wave radiation of sperm and an entering into a nucleus of the quantum of free energy. After radiation the sperm DNA enters into the S-period. The sets of polytene chromosomes of an ovicell and the doubled chromosomes of a sperm undergo the 1st mitotic division of a zygote and give rise to two free nuclei. Then the next DNA-wave radiation of sperm and the 2nd division of a zygote occurs, which is completed by the formation of the 4 nuclei of a 4-nuclear proembryo. The first two divisions of a zygote present a combination of free-nuclear mitosis and cleavage. The mechanisms of these types of mitosis have been described in detail in chapter 3.

According experimental data the first four divisions of a zygote quickly follow one after another are not accompanied by a replication of DNA and not controlled by a zygote nucleus (Ermakov et al., 1981, 1982). That is the switching on of the new program of the development of the future plant takes place only at the proemryogenesis end. The 3-4 division of proembryo is the combination of two types of mitosis – a cleavage and an activating mitosis. This data testifies that the reproductive cycle does not terminate by the fertilization and the zygote formation. Therefore we will touch on this last part of a reproductive cycle in more detail.

As known, the period of free-nuclear divisions is the characteristic phenomenon for the majority of gymnosperms. During these divisions in Araucaria, Ginkgo and Cycas in the center of an ovicell is formed a spherical mass of free nuclei. The superficial nuclei of this structure are located in regular intervals (Haines and Prakash, 1980). The period of free-nuclear divisions and the appearance of an embryonic initial have finished. Part of nuclei of the free nuclear proembryo from the superficial layer participates in the formation of an embryo, and the others have died. A period of free-nuclear divisions is the stage in which there is an emergence of a new reproductive initial (). In *Pinus* the number of free nuclear divisions is reduced to 2 first divisions of a zygote. The nuclei of a 4-nuclear proembryo are positioned in one layer and characterize the totipotency. They serve, as the embryonic initial for the last stage activation.

The 4-nuclear proembryo undergoes the two divisions of activating mitoses, giving rise to a secondary proembryo or a 16-cellular proembryo (Table 2, see). On the course of these divisions the two tiers of unviable energetically weak cells (a primary proximal tier and suspensor tier) depart to the basal pole. The embryonic tier on the apical pole of a 16-cellular proembryo arises: it is the sporophyte initial. The embryo is completely formed from the 4 cells or from even one cell (in the case of cleavage polyembriony) of this tier. I.e., the cells of 16-cellular embryo, which are located in a horizontal plane, possess the identical ultrastructure and function. The cells in a vertical plane, on the contrary, lose the signs of totipotency and considerably differ by morphology, function and destiny. Therefore during proembryogenesis the 2 transverse divisions of activating mitosis are the 6th and the last activation stage ().

During the 4 divisions of proembryogenesis the corpuscular-wave characteristics in the sets of ♂ and ♀ chromosomes are directed to the same parameters. Therefore in each nucleus of the 4 cells of the sporophyte initial the two haploid sets of homologous chromosomes with the same DNA-particle and DNA-wave meet simultaneously. Since the spins of the ♂ and ♀ chromosomes are directed to opposite sides, they are attracted closely to each other and give the connected condition of homologous chromosomes. As a consequence, during interphase the new type of genome – a bisexual diploid set of chromosomes, arises. Because the nucleotides are the Fermi-particles, a diploid genome is a stable condition of genetic material, which is not destroyed during ontogenesis. After attraction the chromosomes become non-polarized particles. Therefore the secondary proembryo is the stage by definition of sex (⚥).

During the interphase of the 16-cellular proembryo a spin-spin interaction of homologous chromosomes, which originate from different parental gametes, takes place and a huge number of casual combinations of genes is created. It is the 3rd stage of the creation of the hereditary program (⌘). Consequently, the changes of genetic material in a nucleus of a secondary proembryo are the connection within one stage of two different reproductive processes – ⚥ + ⌘.

The scheme of a whole reproductive cycle of *Pinus sylvestris L.* is shown in figure 53. The matrix of a reproductive cycle of a species includes the 14 reproductive stages and 4 types of different reproductive processes, which we mark on the scheme as the special symbols – ⬭, ⌘, ᚠ, ⚥. 5 reproductive stages locate on the male generative line, 6 stages – on the female generative line and 3 stages – after fertilization.

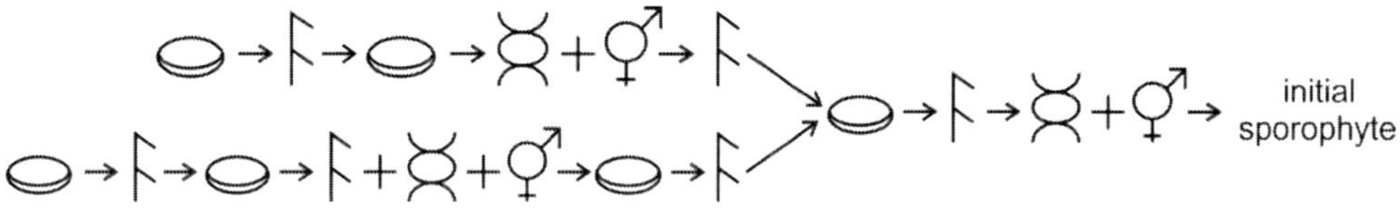

Figure 53. Scheme of reproductive cycle of *Pinus sylvestris* L.

In the end of proembryogenesis the individual and species-specific hereditary (on the basis of DNA-particle) program and the individual and species-specific energy (on the basis of DNA-wave) program of an initial sporophyte are compiled completely. Then an initial sporophyte goes out into the corrosion chamber of endosperm. From the 1st mitotic division of a sporophyte initial the ontogenesis (or the process of independent realization of a hereditary and energy program of plant development, which had been laid in each cell of a sporophyte initial) begins.

The stage of secondary proembryo is long enough (4-5 days). The electron-microscopic studying of proembryogenesis has shown, that the embryonic cells of a 16-cellular proembryo are characterized by the small size and an optimum nucleocytoplasmic ratio. During this stage the changes of ultrastructure, which, possibly, are connected with the formation of their own cytoplasm of an embryo, proceed. According experimental data only an ovicell cytoplasm surrounds a nucleus of a 4-nuclear proembryo, because during cleavage divisions the synthetic processes are absent (Ermakov e. al., 1980, 1981, 1982; Kuznetsova, 1990). Therefore a maternal cytoplasm enters into the cells of an embryo, when it transfers from the free nuclear to the cellular stage of development. During the stage of secondary proembryo the embryo is passed from an ovicell to the corrosion chamber by means of the divisions and the growth of suspensor cells.

The DNA-wave radiation of a new diploid genome of a future plant, replication of DNA and synthetic processes begin at this time as well. Perhaps, during the long interphase the majority of the maternal cytoplasm, which has entered into the embryonic cells of the primary proembryo from a nearby layer of the ovicell (Figure 54-a), lysis. The intensive processes of autophagy and vacuolation are observed (Figure 54-b, c). A great number of primary and secondary lysosomes fill the cytoplasm of the embryonic cells of a 16-cellular proembryo (Figure 54-c). Simultaneously under nucleic control a formation of its own cytoplasm of embryo, which corresponds to a DNA-particle and a DNA-wave of a new genotype of the plant, occurs (Figure 54-d).

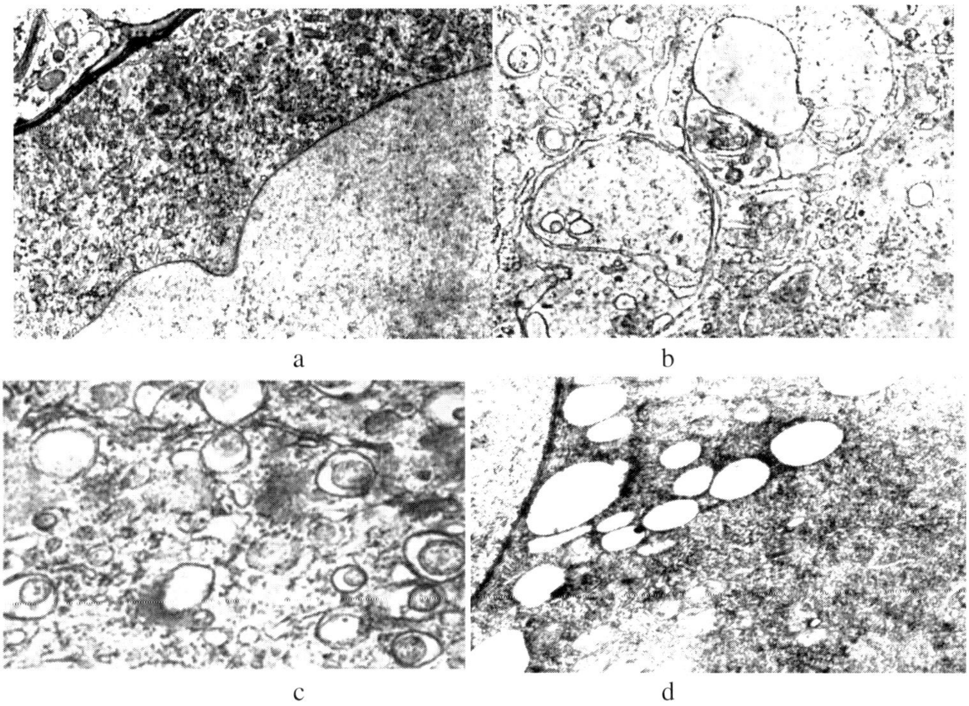

Figure 54. Cytoplasm ultrastructure of 4-nuclear proembryo (a) and embryonic cells of 16-cellular proembryo at the beginning (b, c) and at the end of secondary proembryo stage.

The corpuscular-wave model of a reproductive cycle of all species of higher order plants has a difficult structure. It is possible to present the scheme of the reproductive development of a sporophyte initial in simpler form. The two types of stages participate in the process of "biological assemblage" of a genotype of a future plant – [symbol] and [symbol]. In figure 55 the scheme of a double matrix structure of a reproductive cycle of *Pinus sylveseris* is shown.

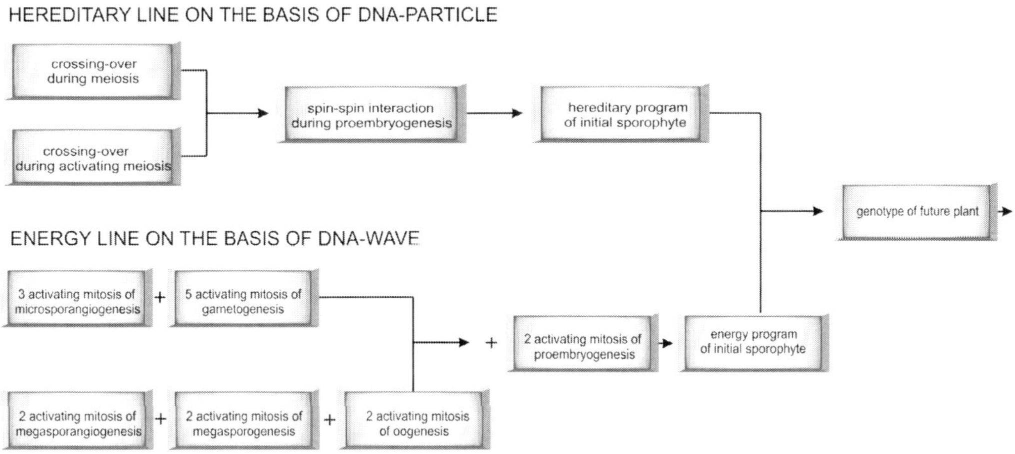

Figure 55. Scheme of structure of the hereditary and energy lines in the matrix of *Pinus sylvestris* L. reproductive cycle.

In this scheme, as was mentioned above, the matrix of a reproductive cycle of pine has a doubled corpuscular-wave nature. I.e., the two main processes – the "biological assemblage" of a hereditary program on the basis of a DNA-particle and an energy program on the basis of a DNA-wave, which present two independent lines in the matrix structure, proceed during the formation of a genotype of a future plant. A hereditary line consists of 3 stages by a DNA-particle change. It is a crossing-over during male and female meiosis and syngamy ♂ and ♀ gametes after fertilization (in the proembryogenesis end). In all higher order plants the number stage in the hereditary line is constant. The *Pinus* energy line consists of 6 activation stages and 16 divisions of activating mitosis. The energy line of plants from the different systematic groups essentially varies on three levels of its organization: 1) a number of activation stages; 2) a inner structure of each activation stage; 3) a number of divisions of cells of a basal pole. As it has been shown in the analysis that the, nature of a species with identical energy lines are absent. During evolution the alterations of a given line have a directed character and testify to the continuous growth of the biological energy of a species.

That hence it follows, that the formation of its own hereditary program during development sporophyte initial takes place on the basis of a DNA-particle. In this prosess the mechanisms of crossing-over (during stages of microsporogenesis, megasporogenesis) and the spin-spin attraction of homologous chromosomes (during stage of secondary proembryo) participate. The microsporangiogenesis, gametogenesis (microgametophytogenesis), megasporangiogenesis, megasporogenesis, oogenesis and proembryogenesis are the stages by the formation in each sporophyte initial of its own energy program (on the basis of a DNA-wave). The formation on one plant of a great number of genetically and energetically non-identical sporophyte initials is the way by which a multiplication of the quantity of an individual of one species. As it has been shown the realization of these three ways, connected in space and time in the uniform corpuscular-wave reproductive process, gives a notion about the biophysical essence of the sexual reproduction of the species.

Conclusion

Thus, according to a corpuscular-wave concept the evolution of the sexual process goes in the direction of the growth of the energy potential of a species. Thanks to it a species receives greater independence of the internal medium from the external environment; the possibility of various ecological niches is mastered; a leading position in the plant world by means of the supplanting of previous energetically weak dominants is occupied. The signs, which testify to the increase of an energy potential of a species we will consider on an example of the reproductive cycle of gymnosperms and angiosperms. In this analysis the following works are used (Kordyum, 1978; Essau, 1980; Embryology of Angiosperms, 1984; Origin and evolution ..., 1988; Gitford and Foster, 1987; Bhatnagar and Moitra, 1996; Williams, 2009; etc.).

1. During the formation of the reproductive complex (stage) a DNA-wave before each division of cells (or nuclei) goes down one energy level. So, the loss of biological energy during the stage of a free-nuclear proembryo makes: in *Pinus* - 2, in *Acaucaria* – 5-6, in *Ginkgo* – 8, in *Dioon edute* – 10 levels. Therefore the evolution of sexual reproduction goes in the direction of the cell number reduction of a structure and the decrease of stage number.

2. The apical meristem of activity cessation (a partial, and then the full in angiosperms) in the time of transition in a floral meristem takes place. This is leads to the increase of the energy potential of ♂ and ♀ initial cells, which give rise to the male and female generative lines.

3. The quantity of cells (or nuclei) of the basal pole diminishes. During activation stages (⊱) it leads to the increase of the biological energy of apical cells. In angiosperms, for example, it is the formation of weakly-developed or tenuitegillate nucellus instead of the crassinucellar ones, a reduction of layer number in a microsporangium wall, etc.

From the aforementioned it become obvious why the gymnosperms evolution the stage of free nuclear proembryo has disappeared, and the matrix of reproductive cycle in *Gnetaceae* and *Welwitschiaceae* acquires the following structure (Figure 56).

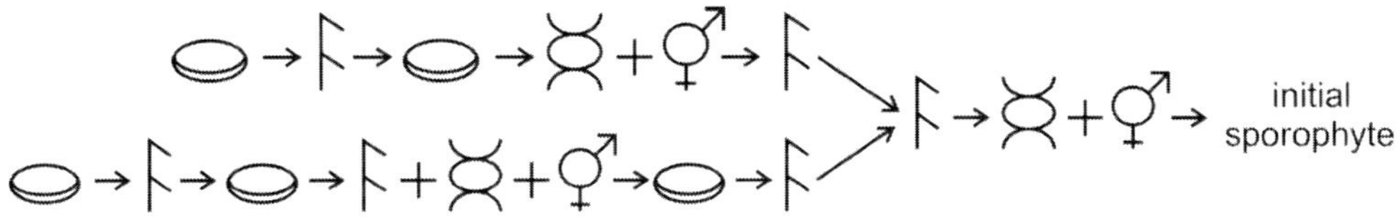

Figure 56. Structure of reproductive cycle in a species of *Gnetaceae* and *Welwitschiaceae* families.

A female gametophyte originates from functional megaspores and during reproductive development passes two activation stages on the basis of a DNA-wave. During evolution of gymnosperms the decrease of cell number in the female gametophyte went on simultaneously with the reduction of stage of the free nuclear proembryo. But this stage, unlike in the previous stage, without the replacement could not disappear from the reproductive cycle. A female gametophyte in the processes of sexual reproduction carries a large functional load. 1. Megagametophytogenesis is the stage of the choice of an archegonium initial. 2. A female gametophyte transfers the energetically strong embryo to the rest state and the boundary of embryonic chamber limits its growth. 3. During embryogenesis and seed germination an endosperm supplies the embryo with nutrients.

The endosperm, as a protective tissue from high biological energy of embryo, possesses the specific reserve capacity. On the one hand the increase of biological energy of species requires the removal the stage of megasporangiogeneis (⬭), because the losses of biological energy occur. On the other hand, this stage has the important role on the processes of energy protection. Therefore the process of speciation on the basis reproductive cycle of gymnosperms began deadlocked. The new structural formation, which is analogical female gametophyte, between energetically strong embryo and the mother plant should appear. In angiosperms, this tissue is the triploid endosperm. The matrix of reproductive cycle of the gymnosperms (or Gnetaceae) is transformed cardinally: the megagametophytogenesis stage disappears, the oogenesis on one division of activating mitosis becomes longer, and two stages after fertilization are reversed. The appearance of the new improved matrix of the reproductive cycle of angiosperms was a consequence of these changes, as it is shown on the figure 57.

The advanced character of angiosperms is a double fertilization. From the new (corpuscular-wave) theoretical positions we will consider the given phenomenon on an example of formation of the most typical embryo sac of the Polygonum-type. The scheme of nuclear divisions during gametophytogenesis is shown in figure 58.

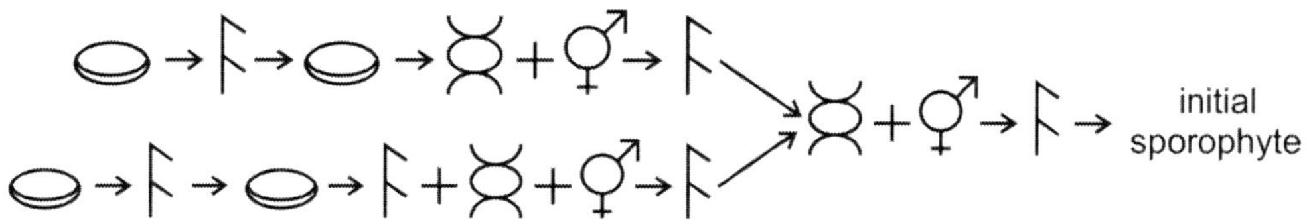

Figure 57. Structure of angiosperms reproductive cycle on the example of *Beta vulgaris* L.

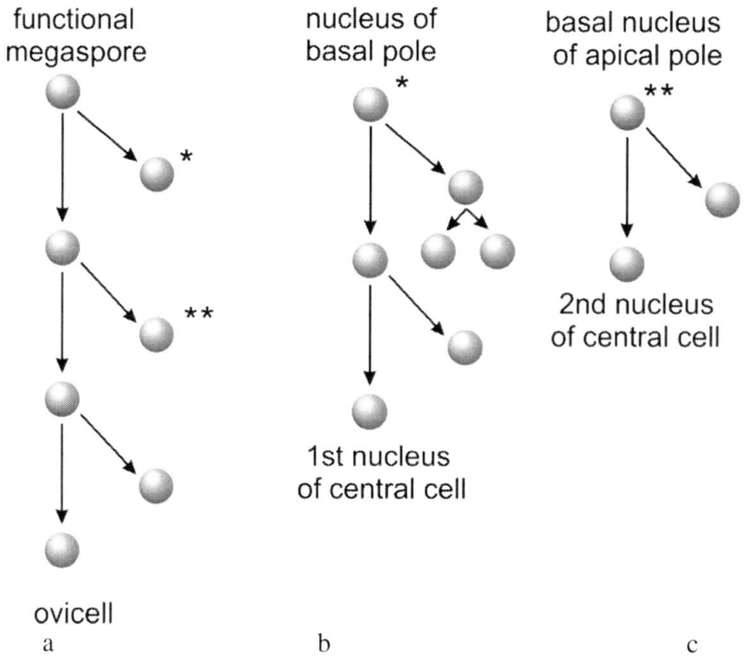

Figure 58. Scheme of cell divisions during development of 8-nuclear embryo sac of Polygonum-type and formation of ovicell (a), 1st (b) and 2nd central cell.

A structure at this stage reminds us of the oogenesis of gymnosperms, but with the length of a reproductive chain in 3 divisions of activating mitosis. In the beginning we will analyze the destiny of the nuclei, which arise during gametophytogenesis, especially, during transverse axial divisions between an initial cell and the apical cell (Figure 58-a). The 1st nucleus division of a functional (chalazal) megaspore is the beginning of an embryo sac formation. In angiosperms it represents the combination of two types of division – a free-nuclear mitosis and an activating mitosis. This division leads to the formation of two nuclei: the more energetically strong apical nucleus – the precursor of egg apparatus, and the less energetically strong basal nucleus – the precursor of antipodal apparatus. The two-nuclear embryo sac grows quickly. An apical nucleus undergoes the 2nd activating mitosis, during which the two energetically non-identical nuclei of the apical and basal pole arise. A 3rd division of an energetically strong nucleus of the apical pole is the last activating mitosis of gametophytogenesis, which ends by the formation of an energetically strong ovicell and an energetically weak cell of a synergid. The nucleus of the basal poles undergoes the two divisions of activating mitosis (Figure 58-b). During these divisions the energetically strong nucleus the 1st central cell and 3 energetically weak cells of antipodes are formed. The 2nd basal nucleus by means of activating mitosis is divided as well (Figure 58-c). Two nuclei – an energetically strong central cell and an energetically weak synergid on the apical pole of an embryo sac have appeared. Soon 5 cells of an embryo sac, which during activating mitosis have lost energy (2 synergids and 3 antipods), without or after a few divisions die. An ovicell, as the most energetically strong generative cell gives rise to an embryo. The two last energetically strong nuclei move to the center of embryo sac where their syngamy occurs.

An egg apparatus of a mature embryo sac consists of two synergids and an ovicell, which is positioned between synergids. During maturation between an ovicell and a sperms, as between the two most energetically strong generative cells, the forces of a spin-spin attraction operate. Therefore the nucleus of an ovicell is moved to an apical part of the cell, and in the synergids the opposite dislocation of nuclei takes place. Two nuclei of the future central cell possess less energy therefore they occupy other spatial positions in a mature embryo sac. They move to the center of a female gametophyte or into the zone, where the forces of spin-spin repulsion with an ovicell and the forces of a spin-spin attraction with a sperm are counterbalanced.

The content of the pollen tube penetrates into the embryo sac through one of the synergids, destroying it, and gets into the zone between an ovicell and the central cell. The nucleus or most energetically strong sperm copulates with a nucleus of an ovicell by the principle ♂↓ + ♀↑ or ♂↑ + ♀↓. A fertilized ovicell becomes the zygote. The nucleus of a less energetically strong sperm passes between the polar nuclei (or cells) of an embryo sac, the fusion of nuclei occurs, giving rise to the central cell (Batigina, 1987). From the Pauli exclusionary principle the connected state of 3 sexual nuclei arises in this case, if their spins are orientated as: ♀↑ + ♂↓ + ♀↑ or ♀↓ + ♂↑ + ♀↓. The product of a threefold fusion of nuclei in the central cell is a primary nucleus of an endosperm.

Thus, during the sexual reproduction of angiosperms a new structural formation – the triploid endosperm begins to carry out the energy protection function. The central cell for an increase in the energy potential has the own way on the basis of 5 activation stages. In spite of the fact that the reproductive cycle consists of a smaller number of stages, the capacity of energy line of initial sporophyte becomes higher.

During transition from gymnosperms to angiosperms one more modification of a reproductive cycle occurs. The capacity of a female energy line becomes higher (7 instead of 6 activating mitosis), and a male generative line, on the contrary, becomes lower (5 instead of 8: the gametogenesis is shortened to 2 activating mitoses). According to the experimental data after fertilization the zygote has a specific period of "resting" and a generative cell in this moment undergoes deep morphological changes (Jensen, 1973; Poddubnaya-Arnoldi, 1976; Batigina, 1987). Apparently, in angiosperms the gametes only with the same DNA-particles and DNA-wave participate in fertilization. Therefore a spin-spin attraction of homologous ♂ and ♀ chromosomes and the formation of a stable diploid chromosome complement (⚥ + stage) occur after syngamy of the nuclei and during this period a zygote is "resting". Accordingly, the preparatory processes for the fusion of ovicell nuclei require more time, than for the fusion of the nuclei of the central cell. The stimulus of zygote division is the 1st DNA-wave radiation of a diploid set of chromosomes. The 1st division of a zygote is the activating mitosis. It is the last activation stage of a reproductive cycle of angiosperms. The apical cell of this division is the sporophyte initial, which during ontogenesis gives rise to a species-specific embryo and plant.

Thus, during the evolution the new type of matrix – a reproductive cycle of angiosperms (Figure 57) has been created. The further process of speciation has proceeded on the basis of this new matrix and till now continues to go in a direction of growth of the biological energy of a species.

References

Allen, G.S. (1946). The origin of the microsporangium of *Pseudotsuga* Bull. *Torrey Bot. Club.* 73, 547-556.

Barantseva,L.M. (1992).*Cytophysiological analysis of archegonium development in Pinus sibirica* Du Tour.Moscow, Moscow State University.

Batigina, T.B. (1987). Some questions of plant embryology. *General embryology.* 4th ed., pp. 440-470, Moscow, Visshaya shkola.

Belajeff, W. (1893). Zur Lehre von dem Pollenschlauche der Gymnospermen. *Ibid.* 11, 196-201.

Belousov, L.V. (2005). *Bases of general embryology.* Moscow, Moscow State University.

Bhatnagar, S.P. and Moitra, A. (1996). *Gymnosperms.* New Age Int., New Delhi. Pvt Ltd Publ.

Bogdanov, Yu. F. (1991). The homologous conjugation of chromosomes. *Improvements of recent genetics.* pp. 61-134. Moscow, Nauka.

Bhatnagar, S.P. and Moitra, A. (1996). Gymnosperms. New Age Int., New Delhi., Pvt Ltd Publ.

Brunkener, L. (1973). Beitrage zur kenntrisder fruben Sporangieenentwicklung der Pteridophyten und der Gymnospermen. *Sven. bot. tidskr.,* 67, 333-400.

Chamberlain, C.J. (1935). *Gymnosperms: structure and evolution.* Chicago, Univ. Chicago Press.

Chromova, L.V. (1985). Embryological processes in a non-pollinated ovicells of pine and anomalies at the xenogamy. *Russ. J. For.*, 2, 47 – 52.

Coulter, J.M. and Chamberlain, C. J. (1917). *Morphology of gymnosperms.* Chicago, The University of Chicago Press.

Dickinson, H.C. (1990). Self-incompatibility in flowering plants. *Bioassays.*, 12, 155-161.

Dickinson, H.G. and Bell, P.R. (!976). Development of the tapetum in *Pinus banksiana* preceding sporogenesis. *Ann. Bot.* 40(165), 103-113.

Drubin, D.G. and Nelson, W.J. (1996). Origin of cell polarity. *Cell, 84,* 335-344.

Dure, A.F. (1976). The visible events of mitotic cell division. *Cell division in higher plants.,* Yaoman M ed., New York, Acad. Press, 49-110.

Embryology of Angiosperms. (1984). Ed. B.M. Johri. Springer – Verlag. Berlin, Heidelborg, New York, Tokye.

Ermakov, I.P., Matveeva, N.P. and Barantseva, L.M. (1980). DNA composition into an ovicell of *Pinus sibirica* Du Tour. on different stages of it development. *DAN SSSR,* 251(1), 254-256.

Ermakov, I.P., Barantseva, L.M. and Matveeva, N.P. (1981). Cytochemical study of DNA in the process of ovicell maturation and early embryogenesis of *Pinus sibirica* Du Tour. *Rus. J. Ontogenesis, 12(4),* 339-345.

Ermakov, I.P., Barantseva, L.M. and Matveeva, N.P. (1982). The structure of the nucleus of developing archegonium and proembryo of *Pinus sibirica* Du Tour. *Rus. J. Ontogenesis, 13(6),* 611-620.

Erspamer, J.L. (1952). *Ontogeny and morphology of the microsporangia in specific genera of the Coniferales.* Diss. Univ. Calif., Berkeley.

Essau, K. (1980). Anatomy of seed plants. Moscow, Mir.

Fagerlind, F. (1961). The initiation and early development of the sporangium in vascular plants. *Sven bot. tidskr.,* 55, 299-312.

Feynman, R.P., Leighton, R.B. and Sands, M. (1963). *The Feynman Lectures of Physics.* vol. 1-3. Wesley, A. (Ed.), Readigg, Massachusetts, Palo, Alto, London.

Fowler, J. and Quatrano, R.S. Cell polarity, asymmetric division and cell fate determination in brown algae zygotes. *Seminar Dev. Biol.,* 6, 347-358.

Fergusson, M.C. (1904). Contribution to the knowledge of the life history of *Pinus* with special reference to sporogenesis, the development of the gametophytes and fertilization. *Proc. Wash. Acad. Sci., 6, 1-202.*

Gilbert, S.F. (2010). *Developmental biology.* 9th ed., Sinauer Associates, Inc., Sunderland, Massachusetts.

Gitford, E.M. and Foster, A.S. (1987). *Morphology and evolution of vascular plants.* 3rd ed. Freeman and Co., New York.

Gorozchankin, I.N. (1923). About corpuscles and sexual process in gymnosperm plants. *Classics of natural sciences.* Moscow, Ptr.

Haines, R.J. and Prakash, N. (1980). Proembryo development and suspensor elongation in *Araucaria* Juss. Ldstr. *J. Bot. (USA). 28, 511-522.*

Herero, M. (2001). Ovary signals for directional pollen tube growth. *Sex Plant Reprod.* 14(1-2), 3-7.

Heslop-Harrison, J. (1978). Genetics and physiology of angiosperm incompatibility systems. *Proc. E. Soc. London.* Ser. B, 202, 73-92.

Hofmeister, W. (1848). *Über die Entwicklung des Pollen.* Bot. Zeitung, 6.

Hofmeister, W. (1851). *Vergleichende Untersuchungen höherer Kryptogamen und der Samenbildung der Coniferen.* Leipzig.

Jensen, W.A. (1973). Fertilization in flowering plants. *BioScience*, 23, 21-27.

Juel, H.O. (1900). Beiträge zur Kenntniss der Tetradentherlung. *Jb. Wiss. Bot., 35(15-16),* 626-638.

Konar, R.N. and Oberoy, Y.P. (1969). Recent work on reproductive structures of living conifers and taxads – a review. *Bot. Rev.* 35(2), 89-117.

Konar, R.N. and Moitra, A. (1980). Ultrastructure, cyto- and histochemistry of female gametophyte of gymnosperms. *Gamete Res.,* 3, 67-97.

Kordyum, E.L. (1978). Evolutionary cytoembryology of angiosperms. Kiev, Nauka.

Kozubov, G..M. (1974). Biology of conifers reproduction in the North. Leningrad, Nauka.

Kozubov, G..M., Trenin, V.V., Tichova M.M. and Kondratjeva, V.P. (1982). The reproductive structure of gymnosperms (a comparative description). Leningrad, Nauka.

Kuznetsova, N.F. (1990). Ulrastructure of female gametophyte of Scots pine. *T cytology,* 32(12), 1168-1174.

Kuznetsova, N.F. (1996). Incompartibility and stages of its display in Scots pine pollination. *Rus. J. For.,* 5, 27-33.

Kuznetsova, N.F. (1998). Activation in the course of reproductive development in female generative sphere of *Pinus sylvestris* L. *T cytology*, 40(10), 841-854.

Kuznetsova, N.F. (2002). The reproductive development on initial sporophyte of *Pinus sylvestris* L. *T cytology*, 44(2), 159-165.

Meier, K.I. (1947). *Systematic of archegonial plants.* Moscow, Leningrad, Sovetskay nauka.

Nekrasova, T.P. (1983). Pollen and pollination regime of Siberia conifers. Novosibirsk, Nauka SO AN SSSR.

Origin and evolution of Gymnospermae. (1988). Ed. Charles B. Beck, Columbia Univer. Press, New York.

Owens, J.N. and Bruns, D. (2000). Western white pine (Pinus monticola) reproduction: Gametophyte development. *Sex Plant Reprod.,* 13, 75-84.

Owens, J.N. and Molder, M. (1971). Meiosis in conifers: prolonged pachytene and diffuse diplotene stages. *Canad. J. Bot.* 49(11), 2061-2064.

Owens, J.N. and Molder, M. (1979). Bud development in *Larix occidentalis.* II. Cone differentiation and early development. *Can. J. Bot.,* 57, 1557-1572.

Owens, J.N. and Molder, M. (1997). Seed-cone differentiation and sexual reproduction in western pine (*Pinus monticola*). *Can. J. Bot.,* 55, 25-2590.

Owens, J.N. and Bruns, D. (2000). Western white pine (Pinus monticola) reproduction: Gametophyte development // *Sex Plant Reprod.* 13, 75-84.

Pandey, K.K. (1980). Evolution of incompatibility systems in plants: origin of "independent" and "complementary"control of incompatibility in angiosperms. *New Phytologist,* 84(2), 381-400.

Pennel, R.I. and Bell, P.R. (1985). Microsporogenesis in *Taxus baccata* L. The development of the archaesporium. *Ann. Bot. (USA)*, 56, 415-427.

Poddubnaya-Arnoldi, V.A. (1976). *Cytoembryology of angiosperms plant. Principles and perspectives.* Moscow, Nauka.

Sarvas, R. (1962). Investigations of the flowering and seed crop of *Pinus sylvestris Comm. Inst. For. Fenn.,* 53, 1- 198.

Schnarf, K. (1933). *Embryologie der Gymnospermen.* Handbuh der Pflanzenanatomie, 11, Berlin.

Singh H. (1978). *Embryology of Gymnosperms.*Gebrüder Borntraeger, Berlin, Stuttgart.

Sinnott, E.W. (1960). *Plant morphogenesis.* New York, McGraw-Hill.

Scupchenko, V.B (1985). *Organogenesis of vegetative and reproductive structures of picea.* Leningrad, Nauka.

Stephens, C., Jenal, U. and Shapiro, L. (1995). Expression of cell polarity during *Caulobater* differentiation. *Seminars Dev. Biol.,* 6, 3-11.

Strasburger, E. (1872). *Die Coniferen und die Gnetaceen.* Jena.

Strasburger, E. (1897). *Angiospermen und Gymnospermen.* Jena.

Trenin, V.V. (1986). *Cytoembryology of Larix.* Leningrad, Nauka

Trenin, V.V. (1988). *Introduction in cytoembryology of conifers.* Petrozavodsk, The Karelian philial AN SSSR.

Tretayakova, I.N. (1990). *Conifer Embryology.* Novosibirsk, Nauka.

Van Vent, J. and Williams, M.T. (1990). Fertilization. *Embryology of plants:use in genetics, selections, biotechnologies.* V. 1. I.P. Ermakov ed. Moscow, Agropromizdat, 224 - 316.

Williams, C.G.(2009). *Conifer Reproductive Biology.* Sprihger Science + Business Media B.V.

List of Terms

Activating mitosis	- is the manner to boost the energy state ("rejuvenation") of generative cells through the transfer of energy from chromosome to chromosome during a special type of cell division. Activating mitosis is ended by the formation of the daughter cells, which have the identical DNA-particle and different energy of DNA-wave. The cell, which has moved to higher levels of energy, begins a new cycle of ontogenesis, or passes on to the next reproductive stage; and the other cell, which has lost the energy, dies (without or after several divisions).
Biological energy	- is the energy of cell as the quantum-mechanical system. Each state of cell corresponds the one of the distinct energy levels, where n = 0, 1, 2, ...
Biofield	- is the electromagnetic field of a cell with two zones: the close-range (changing as $1/r^2$, $1/r^3$ ...) and far-range (changing as 1/r) ones. An obligatory structural element of prokaryotic and eukaryotic cells (during mitosis, in resting and dead cells is absent). The biofield is created by a DNA-wave, because each nucleotide represents the molecular dipole.
Biological harmonic oscillator	- is a quantum harmonic oscillator of all cell systems: in eukaryotic cell as an oscillator serves as the standing DNA-wave oscillating inside a nucleus cavity; in prokaryotic cell - the standing DNA-wave oscillating inside a cell cavity.
Cell system of Stern-Gerlach	- is the three consecutive cell divisions of a reproductive cycle – mitosis and two divisions of activating meiosis. In higher order plants the filtration of the female gametes depending on spin direction takes place on the basis of the Stern-Gerlach cell system. In dioecious plants there are 50% of ovicells with spin-up and 50% of ovicells with spin-down on the exit from this system are formed. In monoecious plants 100% ovicell with a spin-up or with a spin-down (depending on species) arise.

(Continued)

Centrifugal energy stream	- is the path of energy transfer inside the interphase cell during processing. The processing or exponential decay of RNA molecules proceeds with the excretion of free energy. In all prokaryotic and eukaryotic cells takes place the functioning of the given streams are participating only in a single source of internal energy: in prokaryotes - the energy of a DNA-wave, in eukaryotes - the energy of metaphase chromosomes.
Centripetal energy stream	- is the path of energy transfer inside an interphase cell, which operates in the system of "ATP-ADP."In all prokaryotic and eukaryotic cells takes place. All energy sources are involved in the action of this stream: the two of internal (energy of DNA-wave and metaphase chromosomes) and the two of external (energy of photosynthesis and respiration).
DNA-wave energy or 1st source of inner energy	- is the energy of a biological harmonic oscillator, which enters into a cell during a DNA-wave radiation. It is the indispensable structural element of all living cells. In proliferating cells initiates the process of DNA replication and cell doubling. In multicellular eukaryotes, during sexual reproduction the 1st inside source is charged up to a maximum energy level and during ontogenesis it is discharged gradually.
Energetically strong cell and energetically weak cell	- is the notion, which approximately estimates an energy potential of two cells (generative, somatic). The generative cells and their derivatives, which during activating mitosis pass to higher energy levels, become energetically strong ones; and the somatic or generative cells, which have remained at the same level, accordingly, will be energetically weak ones. The gradient arising between them is biological energy, is genetically determined and during cell divisions does not change.
Energy line of reproductive cycle	- is a genetically determined path of the transfer of a generative cell on to the higher levels of energy. The energy line has a species-specific structure and consists of two types of reproductive stages: 1) stages by the definition of a reproductive initial and 2) an activation stage.
Energy protection	- is the different methods and techniques for the protection of a maternal organism from the high biological energy of generative cells. The mechanisms of energy protection began to develop after transition of living matter on a multicellular level, an occurrence of sexual reproduction and during evolution this system has undergone the correct number of cardinal transformations.
Equilibrium cell	- is a cell with the least energy. A DNA-wave is on the zero energy level, or in the basic (unexcited) state. The given level in the ontogenesis of eukaryotic cells is the last, followed by their death. In prokaryotes, the zero level is reversible and the cell for the majority of the cell cycle is in an equilibrium state.

External coiling of chromosomes	- is the manner of chromosome spiralization on the basis of complex isomers during prophase and metaphase mitosis. The external type of coiling leads to the loss of chromosome shoulders of dipole construction and stronger shortening.
Internal coiling of chromosomes	- is the manner of spiralization of a doubled DNA-particle during G_2 period by means of the one-sided nucleotide turnings, the formation of complex isomer links and the gradually shortening of DNA threads.
Isolating layer	- is the boundary layer of extracellular substances (callose, sporpollenin, etc.), which is formed between energetically strong and energetically weak vegetative and generative cells in the zone of their contact. During this process of thickening the isolating envelope constrains, and then stops the growth of reproductive structures on a maternal plant and generative cells pass into a resting state.
Metaphase chromosome energy or 2nd source of inner energy	- is the elastic energy of metaphase chromosomes. The energy is accumulated during chromosome coiling (G_2-period, prophase and metaphase), is utilized by daughter cells during chromosome despiralization (telophase) and is used for the formation of centrifugal and centripetal energy streams, cell growth, etc. (interphase). This inner source only meets in proliferating eukaryotic cells.
Promitosis	- is a variety of mitosis, the manner of cell division of all prokaryotic cells
Sporophyte initial	- is the last generative cell or the first somatic cell of higher order plants. Every initial of a sporophyte possesses the two programs of future plant development - on the basis of a DNA-particle and on the basis of a DNA-wave. In dioecious plants the genetic material of sporophyte initials is differentiated also by sex.

Summary

Quantum biology is one of the most quickly developing directions of modern biology. The novelty of the given corpuscular-wave conception is based on two theoretical statements – the corpuscular-wave nature of DNA and the quantum-mechanical nature of a cell. The wave properties of a plant cell are analyzed mainly at the junction of four sciences – cellular biology, molecular biology, developmental biology and quantum physics (with a minimum of mathematics), by means of the connection of the quantum theory with the experimental data. The elementary and indivisible unit of wave properties of a cell is defined: it is a biological harmonious oscillator – a nucleus (a cell – in prokaryotes) the nucleus together with the oscillating DNA (Ch. 1). A notion is given about the eukaryotic cell as a quantum-mechanical system, an equilibrium system, an opened and closed system, whose development completely submits to the fundamental laws of physics, including the laws of thermodynamics. The corpuscular-wave models of a cell cycle of plant cell (Ch. 2); 9 types of cell divisions (mitosis, asymmetric mitosis, activating mitosis, meiosis, activating meiosis, free-nuclear mitosis, endomitosis, cleavage and promitosis, Ch. 3); a reproductive cycle of higher order plants on the example of *Pinus sylvesteris* L. (Ch. 7) are discussed. It is shown, that a cell development is carried out by the principle of a wave, a spiral, a circle, a line or a point depending on the type of cell system (eukaryotic or prokaryotic) and its vital state (proliferous, differentiated, equilibrium, resting, dead cell, Ch. 4). The supposition is made, that the nature of sexual distinctions of gametes is connected with the mechanism of meiosis and represents the quantum-mechanical phenomenon (Ch. 5). The way of evolutionary development of the sex-linked traits of plants is tracked. For the first time a system of energy protection of plants (the mechanisms, ways and reasons of their change during evolution, Ch. 6) is created. The given system has the wave nature and during interaction of generative and vegetative spheres regulates their relations. The questions of the influence of the fundamental laws of physics on the organization and development of the living matter at different hierarchical levels are discussed. The concept of a two-plane corpuscular-wave (on the basis of a DNA-particle and a DNA-wave) organization of living matter is confirmed by numerous original and literary experimental data and exposes new outlooks for the development of fundamental and practical biology.

Index

A

B

D

E

F

G

H

I

K

L

M

N

O

P

Q

R

S

T

U

V

W

X

Y

Z